Understanding Quantum Mechanics

Detlef Dürr • Dustin Lazarovici

Understanding Quantum Mechanics

The World According to Modern Quantum Foundations

 Springer

Detlef Dürr
Mathematisches Institut
Universität München
München, Germany

Dustin Lazarovici
Faculté des lettres Section de philosophie
Université de Lausanne
Lausanne, Switzerland

Translation by the authors, editing by Stephen Lyle.

ISBN 978-3-030-40067-5   ISBN 978-3-030-40068-2   (eBook)
https://doi.org/10.1007/978-3-030-40068-2

This Springer imprint is published by the registered company Springer Nature Switzerland AG.
The registered company address is: Gewerbestrasse 11, 6330 Cham, Switzerland

Preface

> It needs must be that what can be spoken and thought is; for it is possible for it to be, and it is not possible for what is nothing to be. This is what I bid thee ponder. I hold thee back from this first way of inquiry, and from this other also, upon which mortals knowing naught wander two-faced; for helplessness guides the wandering thought in their breasts, so that they are borne along stupefied like men deaf and blind. Undiscerning crowds, who hold that it is and is not the same, and all things travel in opposite directions! For this shall never be proved, that the things that are not are; and do thou restrain thy thought from this way of inquiry.
>
> Parmenides of Elea, *The Poem, The Way of Truth*[1]

This is not another textbook for a course on quantum mechanics since there are plenty of those. This is a book about the foundations of quantum mechanics that can be read as a companion or supplement to a lecture course or self-study.

The desire, slumbering in us humans, to gain insight into the laws of the cosmos has moved us to write this book. When one decides to study physics as a young person, the decision often comes from this yearning curiosity to understand "was die Welt im Innersten zusammenhält".[2] (Yes, that is quoted from Goethe's *Faust*, who had similar desires.)

But quantum mechanics occupies a special position among all physical theories. There is a plethora of so-called interpretations of its mathematical formalism, which in courses are either barely dealt with or dealt with as briefly as possible in order to get to the heart of the matter quickly: doing calculations like there is no tomorrow. The important and legitimate questions that students are struggling with are rarely addressed at all. Here are some of them:

- What exactly is the role of the wave function in the theory? Is the wave function as real as physical fields are considered to be, for instance, is it as real as the electromagnetic field? Or is it just a mathematical expression of our inability to say what is really going on in nature because our access to the microscopic world (and thus, come to think of it, to the macroscopic world as well) is inherently limited?

---

[1] In: J. Burnet, Early Greek Philosophy, Chap. IV, A. and C. Black, London and Edinburgh, 1930.

[2] Translation by authors: "what holds the world together at its core".

- What exactly is the import of Born's rule (also known as Born's statistical interpretation of the wave function) and where does this rule come from? Obviously, it has something to do with probabilities in measurement experiments that can be carried out in a laboratory. But what really happens, physically, in such a measurement process? How does an electron "produce" a particular measurement result? What laws govern the pointers of a measurement device, and what are these pointers actually made of—all wave function, or what?
- How should we understand the statement that observable quantities in quantum mechanics are described by abstract operators, so-called observable operators? And what about all the sweeping claims that any "realistic" or "deterministic" or somehow more "intuitive" description of quantum phenomena has been proven to be impossible?
- What exactly is the revolution brought about by quantum mechanics? And what about Schrödinger's cat? Is it a manifestation of nature's craziness or is it just a little prank pulled by a brilliant physicist on his confused contemporaries?

Such questions must have clear answers and we have to know those answers if we are to achieve a firm understanding of quantum mechanics.

Sometimes quantum mechanics is still taught in the spirit of the old Copenhagen interpretation that divides the world into two parts: the macroscopic world is classical, the microcosmos is quantum mechanical. The macroscopic world is understandable; the quantum world is not—at least according to the dogma. The two worlds are basically separated by a cut, often called the "Heisenberg cut". There are countless papers on where exactly the cut is to be located, i.e., where exactly the dividing line runs between the microscopic (quantum mechanical) and the macroscopic (classical) world. It is for this reason that the Heisenberg cut is sometimes referred to as the "shifty split".

In this twilight zone, even the wave function itself—the central mathematical object of the theory—becomes obscure. On the one hand, it is supposed to describe a state that somehow brings about measurement results, although these results are random. On the other hand, the "observer"[3]—a subject living in the classical world—uses this object to calculate probabilities of the outcomes of measurements.

We will not say much about the Copenhagen interpretation in this book. Why? Because it just does not make sense. How to see that? The fact that every object in the world consists of atoms (and even smaller units), and atoms are described quantum mechanically. So there can be no such division in principle into classical and quantum mechanical worlds. There is only one world in quantum mechanics, although it can often (usually) be described classically. We shall explain how that is possible.

---

[3]"What exactly is an observer?" is a question that has been discussed way too much. Can a cat be an observer? Or a fly? Or a bachelor student? In his article "Fixing the shifty split" [Physics Today **65** (7), 8 (2012)], Mermin insists that it certainly could not be a mouse.

We can also put it succinctly like this: the Copenhagen interpretation of quantum mechanics is not physics, but a mixture of a mathematical formalism that successfully describes measurement statistics, and psychological warfare against a better understanding.[4]

The following quote regarding the early years of quantum theory takes a hard line on the orthodoxy:

> In the new, post-1925 quantum theory the 'anarchist' position became dominant and modern quantum physics, in its 'Copenhagen interpretation', became one of the main standard bearers of philosophical obscurantism. In the new theory Bohr's notorious 'complementarity principle' enthroned [weak] inconsistency as a basic ultimate feature of nature, and merged subjectivist positivism and antilogical dialectic and even ordinary language philosophy into one unholy alliance. After 1925 Bohr and his associates introduced a new and unprecedented lowering of critical standards for scientific theories. This led to a defeat of reason within modern physics and to an anarchist cult of incomprehensible chaos. (Imre Lakatos in: I. Lakatos and A.E. Musgrave (Hg.), *Criticism and the Growth of Knowledge.* Cambridge University Press, 1970, p. 145.)

Somewhat friendlier was the following remark by Erwin Schrödinger (1887–1961), a few months before his death, in a letter to his friend Max Born, serving as a reminder not to blind ourselves to conformity in the name of an alleged orthodoxy:

> You know, Max, I love you and nothing can change that. But I do need for once to wash your head thoroughly. So stand still. The impudence with which you insist time and time again that the Copenhagen interpretation is practically universally accepted, asserting this without reservation, even before an audience of laity – who are completely at your mercy – is almost unforgivable. You know that Einstein was unsatisfied ("I really never understood what complementarity means" he once said) – as was Louis de Broglie, and I, too, not to mention our poor Max von Laue. Since when, by the way, is a scientific thesis going to be decided by majority vote? (You could certainly reply: at least since Newton.) And excuse me for saying this, but it sometimes seems to me as if you people need repeated emphatic statements to strengthen your own confidence, à la: Sieg-Heil-Sieg-Heil-Sieg ... Are you not afraid of the verdict of history? Are you so convinced that the human race will succumb before long to their own folly? (Erwin Schrödinger, letter to Max Born on 10 October 1960. Translated by the authors.)

Nowadays, the Copenhagen school with all its philosophical ballast is only explicitly represented by a few physicists. What has remained is the tendency to raise the incomprehensibility of quantum physics to the level of a principle. Many introductions to the theory are simply content to emphasise how bizarre and counterintuitive quantum mechanics is, which suggests to learners that they should just study the mathematical formalism and refrain from any more searching questions. It is still sometimes claimed that a rational understanding of the world, in objective terms, has been proven to be simply impossible in quantum physics.

---

[4]For an excellent critical examination of the history and the so-called interpretations of quantum mechanics, see J. Bricmont, *Making Sense of Quantum Mechanics*. Springer, 2016.

The premise of this book is exactly the opposite of this postulate of incomprehensibility, taken as a basic feature of quantum physics: there is no reason to apply a lower standard of clarity, precision, and objectivity to the quantum mechanical description of nature than to the so-called classical theories. A rational understanding of quantum mechanics is possible if we are prepared to throw aside old prejudices and romanticisms and start talking about more than just measurement results. A number of great books have been published in recent years by kindred spirits, making us optimistic that the culture of obscurantism in quantum foundations is slowly but surely coming to an end. We mention in particular Jean Bricmont's *Making Sense of Quantum Mechanics* (Springer, 2016), Travis Norsen's *Foundations of Quantum Mechanics: An Exploration of the Physical Meaning of Quantum Theory* (Springer, 2017), and Tim Maudlin's *Philosophy of Physics: Quantum Theory* (Princeton University Press, 2019). While the "reBELLious" quantum theories presented there are also the ones we discuss, we put a stronger focus on a genetic approach to the mathematical concepts used in quantum mechanics, which in physics courses is often left aside in favour of pure abstraction—and mastering pure abstraction is what keeps many curious minds from asking intimidating questions. Substantial parts of our book are therefore concerned with building bridges from the physics to the mathematics and vice versa.

There are essentially three possible approaches to quantum mechanics that embed the well-known and empirically successful measurement formalism in a precise, fundamental theory. John Stewart Bell (1928–1990) called them "quantum theories without observers", not so much because observers do not occur in them, but because these theories develop an objective description of nature in which "measurements" are subject to the same laws of nature as all other physical processes.

These include Bohmian mechanics, named after David Bohm (1917–1992), in which the statistical predictions of quantum mechanics are derived from the microscopic law of motion for point particles. Another is the GRW collapse theory (named after Ghirardi, Rimini, and Weber), in which the Schrödinger equation is supplemented by a stochastic collapse term. And finally, there is the Many Worlds theory which goes back to Hugh Everett (1930–1982) and which aims at an objective description of nature based solely on the Schrödinger equation.

It is common practice to speak about these as three possible "interpretations" of quantum mechanics. But the term "interpretation" is inappropriate. A poem is interpreted if you want to elicit some deeper meaning from the allegorical language. However, physical theories are not formulated in allegories, but with precise mathematical laws, and these are not interpreted, but analysed. So the goal of physics must be to formulate theories that are so clear and precise that any form of interpretation—what was the author trying to say there?—is superfluous.

Not all the important messages in this book can be expressed in words alone. At the end of the day, a precise formulation of quantum mechanics also requires precise mathematics. However, our goal is not to work through the well-known formalism in all its facets or make it even more abstract. Rather, we shall try to reclassify it where possible, expanding it if necessary and explaining how the statistical predictions of quantum mechanics arise from a fundamental microscopic description of nature which can also be fully deterministic (i.e., without invoking the intervention of random events). But we would also like to encourage you: once the physics is clear, i.e., once it is clear how the theory describes our universe, the mathematics is straightforward, and many apparent contradictions are easily dissolved.

Acknowledgements

What's the point of science? Of the many good answers that can be given, we particularly appreciate one: to understand the universe and our role as a human community in it. Of course, we can reflect on this in splendid isolation. But then, there may be insufficient sources of objection and valuable thoughts of others to help us to question our own ideas in a critical way, and we may end up chasing wild geese. On the other hand, we can also let ourselves be carried along by the broad flow of science, trusting that the path of the many will be the right one. However, there are enough examples throughout history where this approach has also failed. We therefore feel attached to neither one nor the other. Our own thoughts have been changed and sharpened in discussions with many critical friends and colleagues, and the book reflects our present understanding as a result of a long and sometimes hotly debated exchange.

Some chapters in the book are very clearly based on writings by and with friends and colleagues, and we take the opportunity here to mention them by name, saying to all the others that we have not forgotten them, but that only lack of space forces us to be brief. This certainly includes all students at all levels who, in our lectures and in their own research, have turned to us with questions and doubts and forced us to say things better. We have learned a lot from them.

Our thanks go in particular to David Albert, Jeff Barrett, Angelo Bassi, Christian Beck, Jean Bricmont, Siddhant Das, Dirk Deckert, Michael Esfeld, the late Gian-Carlo Ghirardi, Günter Hinrichs, Martin Kolb, Reinhard Lang, Matthias Lienert, Tim and Vishnya Maudlin, Peter Pickl, Paula Reichert, Ward Struyve, Stefan Teufel, Antoine Tilloy, Roderich Tumulka, and Lev Vaidman.

From the names (listed above in alphabetical order) two are left out as they deserve special thanks. These are our friends Shelly Goldstein and Nino Zanghì. Without their insights, help, and cooperation, this book could not have been written.

Special thanks also go to our copy-editor Stephen Lyle (the best in this field) who, with great expertise and empathy, not only made this book readable for non-German-speaking readers but also understood how to convey exactly our intended meaning and yet hold on to something of the originality in our linguistic expression! Special thanks are also due to Dr. Lisa Edelhäuser from Springer, who brought the earlier German edition of this book to life, and the same to Angela Lahee, who took responsibility for the English edition and thereby revitalised a long-lasting friendship between Steve and D.D.

München, Germany   Detlef Dürr
Lausanne, Switzerland   Dustin Lazarovici
December 2019

Contents

Some Mathematical Foundations of Quantum Mechanics

1

> Philosophy is written in that great book that always lies before our eyes – I mean the universe. But one cannot understand it if one does not first learn the language and know the signs in which it is written. This language is mathematics, and the signs are triangles, circles, and other geometric figures, without which it is impossible for man to understand a single word of it; without these one is just wandering around in a dark labyrinth.
>
> Galileo Galilei, *Il Saggiatore*[1]

Here we recall some mathematical basics of quantum mechanics about which there is no dispute. These fundamentals are equally relevant for all quantum mechanical theories (which are unfortunately—or rather, mistakenly—often referred to as interpretations). Our selection is also determined by our needs in later chapters.

The understanding of a physical theory can take place on different levels. The physical worldview must first be communicable to anyone with an honest interest in the subject, i.e., technical details must not be relevant at this level. This means that one should not hide one's own lack of clarity about the worldview behind statements like: "The theory can only be understood by people who have studied physics for at least 4 years". But on a deeper level, the expert level, theoretical (and experimental) background knowledge is necessary for anyone who wishes to have a solid foundation on the basis of which they can explore the intricacies of the description of nature more deeply. The first chapter of this book is meant to provide some key features of that foundation and should be read exactly in this spirit. From a technical point of view, it may be the most challenging chapter, but every physics student should work through the following results carefully at least once in her life. For some, this may mean skipping the more difficult derivations on a first reading and returning to them at a later point in their studies. In any case, the reader should not feel intimidated by the mathematics, as we will try to provide enough context and explanation alongside the technical details.

---

[1]G. Galilei, *Il Saggiatore*, Capitolo VI. 1623. [Translation by authors.]

© Springer Nature Switzerland AG 2020
D. Dürr, D. Lazarovici, *Understanding Quantum Mechanics*,
https://doi.org/10.1007/978-3-030-40068-2_1

There are three mathematical pillars upon which quantum mechanics is based, but which have at the same time given rise to the century long debate about the meaning of quantum mechanics. They are succinctly summarised in Remarks 1.1, 1.2, and 1.4. But before presenting these, we begin by briefly discussing a non-mathematical term, whose omission is in fact the true source for much of the debate about quantum mechanics.

1.1   Ontology

There are a cornucopia of books about the meaning of quantum mechanics. In most of them notions such as *mystical, incomprehensible, quantum logical, information, collapse*, and *observer* pop up almost continually. One particular term, however, hardly ever occurs: *ontology*. The ontology[2] of a physical theory specifies what the theory is about. Since in so-called classical physics it is clear at the outset what the physical theory is about—e.g., Newtonian mechanics is about the motions of point particles—there was no need for an extra Greek word to philosophize about the obvious. But if we wish to understand the confusion about quantum theory, we cannot avoid the term. The reason is simply that, in typical presentations, it is unclear what quantum theory is about. And indeed, each so-called interpretation of quantum mechanics tries to develop its own idea.

John Stewart Bell, who will be mentioned on several occasions throughout this book, invented the term *beables*—a neologism derived from "to be" and "able". Beables are to be contrasted with "observables", or observable quantities. To appreciate the difference, note that observation or measurement is actually a complex physical process. Our measuring devices and sensory organs are complex physical systems that are subject to physical laws and which interact with the measured or observed objects. It is therefore nonsense to think of observed quantities as fundamental in the formulation of a physical theory.[3] With the term *beables*, Bell wanted to stress the fact that a precise physical theory should deal with what there is in the world, i.e., it should deal with the subject of our observations or what causes them. In the mathematical formalism of a physical theory there must therefore be some variables that refer to physical entities out there in the world. These can be particles, fields, strings, or *GRW flashes* (which we shall discuss later)—whatever it is that the theory posits as the elementary building blocks of matter. These elementary objects are the *beables*, as postulated in the ontology of the theory. If the ontology is unclear, then it can never be clear what the theory has to say about the world.

We see a table over there. Why? Because there is a table over there. But physical theories are not about tables as elementary objects. Instead we have an atomistic theory of matter and the table is therefore considered to consist of atoms. We

---

[2]From ancient Greek, meaning the study of "that which is".

[3]Compare with the quote from Einstein at the beginning of Chap. 8.

can conceive of a theory of atoms in which atoms are the fundamental ontology, or *beables*. The theory then provides a way to understand the physical properties of the table from the behaviour of its constituent atoms (possibly through their interactions with fields): its shape, its weight, its temperature, its solidity, its electrical conductivity, etc., can all be explained in terms of atoms. Of course, it has long been known that what we call atoms are not elementary at all. They themselves consist of smaller building blocks, and in such a "finer" theory, these even more elementary building blocks would form the ontology. Ontology also stands for what we consider as being physically "real" in our world, and it is indeed a painful process to learn that what we take to be real can change as theory progresses.

We shall use the term ontology from time to time. We need it to understand quantum theory, because the quandary of orthodox quantum theory is caused by a simple dogma: quantum theory must not be about ontology. But then what is it about? That is what the cornucopia of books take it upon themselves to discuss. In the chapters to come we shall show that the quandary evaporates once quantum mechanics is based on a clear ontology. In a famous German poem by Christian Morgenstern, Palmström concluded razor-sharply that what must not be cannot be. And in fact there were and still are many attempts to turn "shan't" into "can't". We shall talk about that, too.

1.2   The Wave Function and Born's Statistical Hypothesis

A central element of quantum mechanics is the wave function of an N-particle system in three-dimensional space, i.e., in $\mathbb{R}^3$ (this is the generally accepted way of speaking even in quantum theories in which particles do not occur as entities at all):

$$\psi : \mathbb{R}^{3N} \times \mathbb{R} \to \mathbb{C}, \quad \psi(\mathbf{q}_1, \ldots, \mathbf{q}_N, t). \qquad (1.1)$$

Here $\mathbb{C}$ is the set of complex numbers, that is, ψ is a complex-valued function which takes as input a time t and N points in $\mathbb{R}^3$, describing a possible configuration of N particles in three-dimensional space. The time evolution of the wave function with potential V obeys the Schrödinger equation, which we write in terms of the *configuration* variable $q = (\mathbf{q}_1, \ldots, \mathbf{q}_N) \in \mathbb{R}^{3N}$:

$$i\hbar \frac{\partial}{\partial t} \psi(q, t) = -\sum_{n=1}^{N} \frac{\hbar^2}{2m} \Delta_n \psi(q, t) + V(q) \psi(q, t), \qquad (1.2)$$

with the Laplace operator $\Delta_n = \partial^2 / \partial \mathbf{q}_n^2$.

There is a disagreement about whether wave functions actually exist for systems of very "large" size, e.g., a measuring apparatus in a laboratory, or the laboratory itself, or even the whole universe. This disagreement will be discussed in Chap. 2 on the measurement problem. However, we mention the origin of the disagreement here because it runs through the whole of quantum mechanics.

Remark 1.1 (Superposition Principle) The Schrödinger equation is a *linear* (partial differential) equation. This means that the sum of constant multiples of solutions of the equation is also a solution of the equation. In the usual jargon, we say that solutions can be superposed.

In addition to the Schrödinger equation, there is a second important equation which led Max Born (1882–1970) to the accepted interpretation of the wave function as a probability amplitude, although only after a correction by Schrödinger.[4] In common parlance, Born's statistical interpretation, which is often referred to as Born's statistical hypothesis or Born's rule, can be stated as follows:

Remark 1.2 (Born's Statistical Hypothesis) If a system has wave function ψ, the measured positions of the particles are distributed according to $\rho = \psi^*\psi = |\psi|^2$. Here ψ^* is the complex function conjugate to ψ.

This means that, if $A \subset \mathbb{R}^{3N}$ is a (measurable[5]) subset of the configuration space, then the probability of finding the system configuration Q in A is given by

$$\mathbb{P}^\psi (Q \in A) = \int_A |\psi|^2(\mathbf{q}_1, \ldots, \mathbf{q}_N) \, d^3q_1 \ldots d^3q_N . \tag{1.3}$$

We note that Born's interpretation gives rise to probabilities in quantum mechanics which then appear in quite different forms. The second equation mentioned above, which is central to this statistical interpretation, is usually derived in textbooks by computing $\partial|\psi|^2/\partial t$ using Schrödinger's equation. The reader is encouraged to carry out this derivation using:

1. the product rule for calculating the derivative of a product,
2. the fact that ψ^* solves the complex conjugated form of the Schrödinger equation (1.2), and
3. the fact that the potential V takes real values, so it drops out in the end.

This leads to a continuity equation, the so-called quantum flux equation:

$$\frac{\partial|\psi|^2}{\partial t} = -\nabla \cdot j^\psi , \tag{1.4}$$

where $\nabla = (\nabla_1, \ldots, \nabla_N)$, $\nabla_k = \partial/\partial\mathbf{q}_k$, and the *quantum flux* $j^\psi = (\mathbf{j}_1^\psi, \ldots, \mathbf{j}_N^\psi)$ is given by

$$\mathbf{j}_n^\psi = \frac{\hbar}{2im}(\psi^*\nabla_n\psi - \psi\nabla_n\psi^*) = \frac{\hbar}{m}\mathrm{Im}\,\psi^*\nabla_n\psi . \tag{1.5}$$

---

[4]Born had first thought of $|\psi|$ as a candidate for a probability density.
[5]In the sense of mathematical measure theory.

Here Im denotes the imaginary part. The usual argument for $\rho = |\psi|^2$ then proceeds as follows. Integrate (1.4) over the entire configuration space $\Gamma = \mathbb{R}^{3N}$, then transform the volume integral into a surface integral on the right-hand side by application of Gauss' theorem, so that the quantum flux gets integrated over a surface $\partial\Gamma$ at infinity, where the flux is zero:

$$\frac{d}{dt}\int_\Gamma \rho\, d^{3N}q = \int_\Gamma \partial_t \rho\, d^{3N}q = -\int_\Gamma \nabla\cdot j^\psi\, d^{3N}q = \int_{\partial\Gamma} j^\psi\cdot d\sigma = 0. \quad (1.6)$$

This shows that the integral $|\psi|^2$ over the whole space is preserved in time and $|\psi|^2$ can indeed be taken as a probability density, because clearly, the total probability, the probability of the sure event, cannot change in time. Normalised to unity, it remains forever at unity. Of course, the invariance of the measure is only a necessary condition, not a sufficient condition, to be able to consider $\rho = |\psi|^2$ as a meaningful probability distribution. In textbook quantum mechanics, the Born rule therefore has the status of a postulate whose setting is ultimately only justified by experiment. However, a theoretical justification of Born's rule is possible, and we shall discuss this in Chap. 4.

1.3   The Spreading of the Wave Packet

An important phenomenon associated with the Schrödinger evolution is the spreading of a wave packet. With some basic mathematical knowledge, it is easily explained. The wave function (in the form of a wave packet) of a particle of mass m can (and should) be thought of as a superposition of plane waves, i.e., in mathematical terms, we should consider its Fourier decomposition. A plane wave with wavelength λ and wave number $k = 2\pi/\lambda$ evolves according to

$$e^{i(\mathbf{k}\cdot\mathbf{x}-\hbar k^2 t/2m)},$$

as can be checked immediately using the "free" Schrödinger equation (1.2) for one particle and for potential $V = 0$. Here $\mathbf{k}$ is the wave vector with length $|\mathbf{k}| = k$. The superposition of the plane waves with weights $\hat{\psi}_0(\mathbf{k})$, i.e., the Fourier transform of $\psi(\mathbf{x}, 0)$, yields

$$\psi(\mathbf{x}, t) = \int \hat{\psi}_0(\mathbf{k})e^{i(\mathbf{k}\cdot\mathbf{x}-\hbar k^2 t/2m)}d^3k, \quad (1.7)$$

whence we may assign a group velocity $\mathbf{v}$ to the wave group around a certain value of $\mathbf{k}$:

$$\mathbf{v} = \frac{\hbar\mathbf{k}}{m}. \quad (1.8)$$

To obtain this, we differentiate the dispersion relation $\omega := \hbar k^2/2m$ with respect to $\mathbf{k}$ and evaluate at that $\mathbf{k}$-value around which the wave group is centered.

The formula (1.8) was found long before the Schrödinger equation by Louis de Broglie (1892–1987) as a generalisation to "matter waves" of Einstein's T-shirt formula for photons $E = h\nu$. Indeed, $\mathbf{p} = \hbar\mathbf{k}$ is de Broglie's relationship between the wave number and momentum of the particle. The shorter the wavelength, the faster the wave moves. That's one thing to keep in mind. On the other hand, it is known from lectures on analysis, and in particular, the Fourier transform, that the more concentrated a function is in space, the more plane waves with higher k-values occur in the Fourier decomposition of the function. So if we consider the wave function of a particle that is highly localised around a position, then we know very exactly the location of the particle due to Born's interpretation. And the more closely we need to know the particle's position, the more localised the wave function must be. However, the more localised the wave function, the more plane waves with ever higher k-values will be needed to compose that wave function. And then, since the plane waves all have different speeds, the original wave will break up in the course of time into the plane wave parts running at different speeds, at least when the free Schrödinger equation with $V = 0$ governs the motion. The location of the particle at time T will thus be widely scattered, because the individual waves have travelled different distances, and all the more so as the spread in k-values increases.

Every student of physics should carry out the mathematical examination of this spreading effect at least once. It is as fundamental as the derivation of the quantum flux, and that is why we discuss it here. To do so, we first read the integral in (1.7) as an inverse Fourier transform of the product of the functions $e^{-i\hbar k^2 t/2m}$ and $\hat{\psi}_0(\mathbf{k})$. Next we recall from the study of analysis that the product of two functions becomes a *convolution* under Fourier transform:

$$\widehat{f \cdot g}(\mathbf{k}) = \hat{f} * \hat{g}(\mathbf{k}) := \frac{1}{(2\pi)^{3/2}} \int \hat{f}(\mathbf{k} - \mathbf{k}')\hat{g}(\mathbf{k}')\,\mathrm{d}^3 k'\,.$$

The same also applies to the inverse transform. Our first mathematical task will thus be to determine this convolution. We compute the Fourier transform of the first function which looks like a Gaussian up to the factor i, noting that the presence of this factor does not change the result that the Fourier transform of a Gaussian is again a Gaussian.

The rigorous calculation is more involved and uses complex analysis. The result is

$$\psi(\mathbf{x}, t) = \int \frac{1}{\left(2\pi\,\mathrm{i}\frac{\hbar}{m}t\right)^{3/2}} \exp\left[\mathrm{i}\frac{(\mathbf{x} - \mathbf{y})^2}{2\frac{\hbar}{m}t}\right] \psi_0(\mathbf{y})\mathrm{d}^3 y\,, \qquad (1.9)$$

which is an important representation of the evolution of the wave function for the initial wave function ψ_0. Evaluating the square in the Gaussian, we obtain

$$\psi(\mathbf{x}, t) = \frac{1}{\left(it\frac{\hbar}{m}\right)^{3/2}} \exp\left(i\frac{\mathbf{x}^2}{2\frac{\hbar}{m}t}\right) \int \frac{1}{(2\pi)^{3/2}} \exp\left(-i\frac{\mathbf{x}\cdot\mathbf{y}}{\frac{\hbar}{m}t}\right) \exp\left(i\frac{\mathbf{y}^2}{2\frac{\hbar}{m}t}\right) \psi_0(\mathbf{y}) \, d^3y$$

$$= \frac{1}{\left(it\frac{\hbar}{m}\right)^{3/2}} \exp\left(i\frac{\mathbf{x}^2}{2\frac{\hbar}{m}t}\right) \hat{\psi}_0\left(\frac{\mathbf{x}m}{t\hbar}\right) \tag{1.10}$$

$$+ \frac{1}{\left(it\frac{\hbar}{m}\right)^{3/2}} \int \frac{1}{(2\pi)^{3/2}} \left[\exp\left(i\frac{\mathbf{y}^2}{2\frac{\hbar}{m}t}\right) - 1\right] \exp\left(-i\frac{\mathbf{x}\cdot\mathbf{y}}{\frac{\hbar}{m}t}\right) \psi_0(\mathbf{y}) \, d^3y \, .$$

Note that the second summand in (1.10) goes to zero[6] when $t \to \infty$, because

$$\lim_{t\to\infty}\left[\exp\left(i\frac{\mathbf{y}^2}{2\frac{\hbar}{m}t}\right) - 1\right] = 0\,.$$

This means that for large times the wave function is given by the first summand in (1.10), viz.,

$$\psi(\mathbf{x}, t) \approx \frac{1}{\left(it\frac{\hbar}{m}\right)^{3/2}} \exp\left(i\frac{\mathbf{x}^2}{2\frac{\hbar}{m}t}\right) \hat{\psi}_0\left(\frac{\mathbf{x}m}{t\hbar}\right)\,. \tag{1.11}$$

This can be interpreted as follows. For large times t the wave function will have moved to places $\mathbf{x}$ for which $\mathbf{k} = \mathbf{x}m/t\hbar \in \text{supp}\,\hat{\psi}_0$, that is, places which are reached by wave groups centered around those wave vector values. Here $\text{supp}\,\hat{\psi}_0$ is the set of values for which $\hat{\psi}_0 \neq 0$, where supp is an abbreviation for "support". Hence, in a sense, $\text{supp}\,\hat{\psi}_0$ specifies which plane waves are contained in ψ_0, and these then diverge from each other at different speeds according to the de Broglie relation $m\mathbf{v} = \hbar\mathbf{k}$ for the momentum.

We also find that the momentum $\hbar\mathbf{k}$ is distributed according to the probability density $\left|\hat{\psi}_0(\mathbf{k})\right|^2$. This can be seen as follows. Suppose $\psi_0(\mathbf{x})$ is concentrated around $\mathbf{x} = 0$. The particles that have reached the position $\mathbf{X}(t)$ at time $t \gg 0$ have therefore moved approximately with the average momentum $\hbar\mathbf{k} = \frac{m}{t}\mathbf{X}(t)$.

---

[6]Strictly speaking, we should use Lebesgue's theorem of dominated convergence here, because we exchange integration with taking a limit, but rigor does not bring new insights, so let's ignore that here.

Hence, for any (measurable) subset $A \subseteq \mathbb{R}^3$, the momentum distribution reads

$$\mathbb{P}^\psi \left(\hbar \mathbf{k} \in \hbar A \right) \approx \mathbb{P}^\psi \left(\mathbf{X}(t) \in \frac{t\hbar}{m} A \right)$$

$$\approx \left(\frac{m}{\hbar t} \right)^3 \int_{\frac{t\hbar}{m} A} \left| \hat{\psi}_0 \left(\frac{\mathbf{x}m}{t\hbar} \right) \right|^2 \mathrm{d}^3 x = \int_A |\hat{\psi}_0(\mathbf{k})|^2 \mathrm{d}^3 k , \qquad (1.12)$$

where we have substituted $\mathbf{k} := \mathbf{x}m/t\hbar$ in the last step. Note that everything follows from Born's rule for the position distribution. And in particular note that we consider here an average momentum, i.e., we know that the particle is at time $t = 0$ here and at a much later time there. Then we take the distance between the positions and divide by the time difference. It is this quantity that is measured during a "momentum measurement". The expected value of the momentum $\mathbf{P} = \hbar \mathbf{k}$ is according to its definition

$$\mathbb{E}^\psi(\mathbf{P}) = \int_{\mathbb{R}^3} \hbar \mathbf{k} \left| \hat{\psi}_0(\mathbf{k}) \right|^2 \mathrm{d}^3 k . \qquad (1.13)$$

This can be rewritten as

$$\mathbb{E}^\psi(\mathbf{P}) = \int_{\mathbb{R}^3} \hat{\psi}_0^*(k) \, \hbar \mathbf{k} \, \hat{\psi}_0(\mathbf{k}) \, \mathrm{d}^3 k . \qquad (1.14)$$

In lectures on quantum mechanics we are told that the momentum observable in the position representation is given by $\frac{\hbar}{i} \nabla$, and we can now easily understand what this means from (1.14):

$$\hbar \mathbf{k} \, \hat{\psi}_0(\mathbf{k}) = \hbar \mathbf{k} \int \psi_0(\mathbf{x}) e^{-i\mathbf{k}\cdot\mathbf{x}} \mathrm{d}^3 x$$

$$= \hbar \int \psi_0(\mathbf{x}) \, i\nabla \, e^{-i\mathbf{k}\cdot\mathbf{x}} \mathrm{d}^3 x$$

$$= \hbar \int \left[-i\nabla\psi_0(\mathbf{x}) \right] e^{-i\mathbf{k}\cdot\mathbf{x}} \mathrm{d}^3 x \qquad \text{(using integration by parts)}$$

$$= \left(\widehat{\frac{\hbar}{i}\nabla\psi_0(\mathbf{x})} \right) \qquad \text{(by Fourier transform)} .$$

By virtue of the Plancherel identity we then get for (1.14)

$$\mathbb{E}^\psi(\mathbf{P}) = \int_{\mathbb{R}^3} \psi_0^*(\mathbf{x}) \frac{\hbar}{i} \nabla \psi_0(\mathbf{x}) \, \mathrm{d}^3 x .$$

This leads to the introduction of the *momentum operator* $\hat{\mathbf{P}} = \frac{\hbar}{i}\nabla$. In "position space" it is a gradient and in "momentum space" a multiplication operator, multiplying the Fourier transform of the wave function by $\hbar\mathbf{k}$.

Remark 1.3 (Perspectives on Heisenberg's Uncertainty Relation) We have seen that the spreading of the wave function is a wave phenomenon which, in combination with Born's statistical hypothesis, leads to an empirical truth which can justifiably be regarded as one of the most significant innovations of quantum mechanics: the more spatially concentrated a wave function, the more quickly it spreads in time. According to Born, the initial high spatial concentration allows a rather exact knowledge of the position, while the spreading leads to a whole range of possible end positions, whose distribution we have associated with the momentum. The relationship between the exactness of position and momentum measurements is described by Heisenberg's uncertainty relation.

Unfortunately, this is often interpreted as meaning that the uncertainty relation implies the impossibility of simultaneous, arbitrarily accurate measurements of momentum and position, but also that it proves that there can be no particles moving on trajectories in quantum mechanics. Such an assertion is of course unwarranted, but as the derivation of the uncertainty relation is usually presented in a highly abstract manner, its true meaning often remains foggy. For example, the mathematically rigorous derivation of the uncertainty relation is based on the commutator of the "position and momentum observables". It is then claimed that the commutator lies at the heart of the uncertainty relation. But before buying that, we need to be clear about the role the "observables" actually play in the theory. We shall clarify this role in Chap. 7.

However, we can already understand from what has been said so far that what is really responsible for the uncertainty relation is the time evolution of Schrödinger's wave function paired with Born's interpretation. And this leads to a different question: how can we justify Born's interpretation? To come to grips with that, we need to understand the so-called quantum theories without observers, such as Bohmian mechanics. In such theories we will explain and justify Born's statistical rule. Finally, we shall understand why the uncertainty relation has nothing to do with the existence or non-existence of particle paths. And the best way to see this is to derive it from a theory in which particles move on trajectories, as we shall do in Sect. 4.3.

1.4 No Mystery: The Double-Slit Experiment

Richard P. Feynman (1918–1988), one of the great physicists of the last century, began his lecture on the double-slit experiment as follows:

> In this chapter we shall tackle immediately the basic element of the mysterious behavior in its most strange form. We choose to examine a phenomenon which is impossible, *absolutely* impossible, to explain in any classical way, and which has in it the heart of quantum

mechanics. In reality, it contains the *only* mystery. We cannot make the mystery go away by "explaining" how it works. We will just tell you how it works. In telling you how it works we will have told you about the basic peculiarities of all quantum mechanics. (R.P. Feynman, R.B. Leighton and M. Sands. Feynman Lectures on Physics, Volume 3, pp. 17–18, http://www.feynmanlectures.caltech.edu.)

Now we could say that if even Feynman felt that the double-slit experiment was a mystery, then we shouldn't be ashamed if we feel the same way. That's one possible angle. However, another point of view is this: if after almost a century standard quantum mechanics is still unable to explain—in the true sense of "explain"—such a fundamental phenomenon, then there may be something wrong with the theory. The quantum theories we will discuss in this book will clarify the double-slit experiment and other quantum phenomena, in the sense that they not only "describe" the phenomena correctly, but provide a complete picture of how the phenomena come about.

But let's stick to the Feynman quote for now. Feynman says that it is absolutely impossible to explain the double-slit experiment "in a classical way". This is intuitively correct if "classical" refers to the laws of Newtonian mechanics. However, in addition to Heisenberg's uncertainty relation, the double-slit experiment is often quoted as proof that "classical logic" has lost its validity in quantum mechanics. Or that the experiment shows that particle trajectories are simply impossible. Both views are wrong. The possible particle trajectories are shown in Chap. 8, but to come to grips with the discussion at hand the reader is advised to look at Fig. 8.2 and focus for the time being only on the end points of the trajectories. They represent the typical blackening on the screen, which results in the famous "interference pattern".

So what is all this talk about the failure of "classical logic"? Actually, it is rather strange that such concepts as "quantum logic" and "classical logic" ever entered scientific discussions at all, in particular since the double-slit experiment has nothing to do with all this. The double slit experiment can be described as follows. Two (nearby) slits are imaged by a particle beam on a photo screen and the blackening of the screen shows an interference pattern. More precisely, at very low beam intensities, i.e., only one particle is on the way[7] at any given time, or put another way, we send only one-particle waves through the slits, the interference pattern, made up of completely localized, randomly distributed black dots, develops slowly over time.[8] This is perhaps the most important message of all: the interference pattern emerges gradually, wave by wave (or particle by particle), and is nothing more than an accumulation of impact points. Once this has been understood, it opens up a whole catalogue of questions, from which we would like to draw attention to two in particular:

---

[7] In an ideal situation, we might imagine sending only one particle through the slits each day ...

[8] ... and that could take years.

1. Why does a pointlike spot appear when it is actually a wave that impinges on the screen? The answer closely concerns the measurement problem of quantum mechanics, which we discuss in Chap. 2 .
2. What is the time distribution of the dots? Suppose the one-particle waves are prepared and sent at fixed times, let's say every full hour. Does every dot appear on the screen after the same amount of time? The answer is no. The times are random, with statistics determined by the quantum flux (1.5). This is usually ignored in textbook presentations and there are various ways to justify that. The main reason why this can be ignored from a practical point of view relates to the asymptotic shape of the wave function (1.11), which ensures that the flow lines eventually become straight. When calculating the distribution of dots on the screen, we can in fact replace the real wave pattern behind the slit by a stationary (i.e., temporally unchanging) wave pattern and apply Born's interpretation to this. The exact analysis is mathematically rather involved, however. It can be found in the chapter on scattering theory in *Bohmian Mechanics, The Physics and Mathematics of Quantum Theory.*[9]

We turn now to the issue of logic. First let slit 1 be closed, then the particle can only go through slit 2 and we can make the following statements about this experiment:

> The particle goes through slit 2 and hits the screen at $\mathbf{x}$. (1.15a)

> The corresponding probability is $|\psi_2|^2(\mathbf{x})$. (1.15b)

Here ψ_2 is the wave function that emerges from slit 2 as a spherical wave. Here we repeat our previous warning: recalling the answer to question (2) above, the true probability for dots appearing on the screen is determined by the quantum flux, but the probability can effectively be computed from the $|\psi_2|^2$ distribution at a fixed (large) time. The same applies to the case when we close slit 2 and open slit 1. In this experiment, we can make the statements:

> The particle passes through slit 1 and hits the screen at $\mathbf{x}$. (1.16a)

> The corresponding probability is $|\psi_1|^2(\mathbf{x})$. (1.16b)

The experiment in which both slits are open can then be described as follows:

> The particle passes through either slit 1 or slit 2 and hits the screen at $\mathbf{x}$. (1.17a)

> The corresponding probability is

$$|\psi_1(\mathbf{x}) + \psi_2(\mathbf{x})|^2 = |\psi_1|^2(\mathbf{x}) + |\psi_2|^2(\mathbf{x}) + 2\mathrm{Re}\,\psi_1^*(\mathbf{x})\psi_2(\mathbf{x}), \quad (1.17b)$$

---

[9]D. Dürr and S. Teufel, *Bohmian Mechanics. The Physics and Mathematics of Quantum Theory.* Springer, 2009.

where Re is the real part and we observe that this is not equal to $|\psi_1|^2(\mathbf{x}) + |\psi_2|^2(\mathbf{x})$. The difference with (1.17b) is due to the interference between the wave function parts ψ_1 and ψ_2 which emerge from slits 1 and 2. This is a typical wave phenomenon. A water wave that passes through a double slit develops the spherical Huygens waves behind the slits, named after Christiaan Huygens (1629–1695), and these then produce the same interference pattern. In the present case the interference term is given by $2\mathrm{Re}\psi_1^*(\mathbf{x})\psi_2(\mathbf{x})$. And this is the bad news, because (1.15a) and (1.16a) are the alternatives in (1.17a), so logically the corresponding probabilities should add up. But they don't!

So we conclude that classical logic fails. Either that or the particle interpretation is nonsense. Or better still, both. It would only be if

$$\mathrm{Re}\,\psi_1^*\psi_2 = 0 \,, \tag{1.18}$$

that is, if the interference vanishes, that the probabilities (1.15b) and (1.16b) would sum and ordinary logic would be redeemed.

But the truth is, this is all a lot of fuss over a red herring. All we need to do is to take physics seriously and describe the two situations correctly:

> Slit 1 (2) is closed and the particle goes through slit 2 (1) and hits the screen at $\mathbf{x}$.

These two statements correspond to the physical situations in the first two experiments, but they are *not* the alternatives in (1.17a). This is obvious, because in the latter case both slits are open. This difference in the physical settings should be noted first. Then we may go on and realize that the assertion that the probabilities associated with the first two experiments should "logically" sum up is based on the assumption that the behaviour of the particles passing through slit 1 does not depend on whether slit 2 is open or closed (and vice versa). The first point is that this assumption is not justified, and the second is that it is clearly wrong in quantum mechanics: if both slits are open, then the wave passing through both slits interferes with itself. The resulting waveform is certainly responsible for the shape of the accumulation of the black dots. The only question that remains is: Where do the black *dots* come from? This is indeed a rather disturbing and yet crucial question because, as already stressed, it is intimately connected with the measurement problem, to be dealt with in Chap. 2.

1.5   The Importance of Configuration Space

After linearity (Remark 1.1) and Born's statistical law (Remark 1.2), we come now to the third pillar of quantum mechanics. For this, we go back to the beginning, namely to the wave function of an N-particle system [see (1.1)]. No matter how abstract or mind-boggling it may sometimes seem to be, the wave function $\psi(\mathbf{q}_1, \ldots, \mathbf{q}_N)$ is first and foremost simply a function on the configuration space of the N particles. Those who believe that the concept of "particle" is physically ill-founded can take this sentence and the word "particle" as a "manner of speaking". A configuration is the collection of N position variables $\mathbf{q}_i \in \mathbb{R}^3$, i.e., $q = (\mathbf{q}_1, \mathbf{q}_2, \ldots, \mathbf{q}_N)$ and the set of all possible configurations is called the configuration space, denoted $\mathbb{R}^{3N}$.

Remark 1.4 (Configuration Space and Entanglement) The wave function of an N-particle system is defined on the configuration space $\mathbb{R}^{3N}$. This means that all N particles share a common wave function. Only in the special case when the wave function is a product $\psi(\mathbf{q}_1, \mathbf{q}_2, \ldots, \mathbf{q}_N) = \varphi_1(\mathbf{q}_1)\varphi_2(\mathbf{q}_2)\cdots\varphi_N(\mathbf{q}_N)$ can we say that the wave function of the whole system is given by N independent wave functions, one for each particle. In quantum mechanics, such a state is said to be "separable". In general, however, $\psi(\mathbf{q}_1, \ldots, \mathbf{q}_N)$ is *not* a product, in which case we speak of quantum *entanglement* or we say the "state is entangled".

The main difficulty with thinking in terms of configuration space is that it has very high dimensions, e.g., for a gas of particles, the dimension is of the order of Avogadro's number ($\sim 10^{24}$). It is therefore hard to picture.[10] On the other hand, we really *must* think in terms of configuration space in order to understand quantum mechanics! In later chapters we shall return often to this point, but for the time being we wish to emphasize the meaning of configuration space and to put forward some ideas that can help us to reason in configurational terms.

Our world of experience is spatially located in a three-dimensional space,[11] i.e., bodies like measurement instruments and their pointers occupy regions of space. But the bodies themselves consist of atoms (which themselves consist of smaller objects, but that doesn't matter now) and it is their spatial arrangement, namely their *configuration* that gives the body its shape. This means that the *region in configuration space* determines the body, its shape, its position, and its orientation in space. What must be understood is that different spatial positions of a body are described by disjoint regions in configuration space. If we are concerned with macroscopic bodies (where the number of atoms is of the order of Avogadro's constant), for instance the pointer of a piece of measuring apparatus, then macroscopically different situations (pointer points to the left or to the right) are given by macroscopically disjoint subsets of configuration space (see Fig. 1.1).

---

[10]The rather beautiful German word "unanschaulich" is often used to describe configuration space, but it has also been abused in the context of quantum mechanics to suggest that one cannot develop any coherent picture of what is going on in the microcosm.

[11]It is often stated in quantum theory that it is of great philosophical importance that in Hilbert space, the space of wave functions, one can select an arbitrary basis in which to express the wave function in coordinates. So there is a position basis, a momentum basis, an energy basis, and whatever we want, and none of those bases is preferred. This is true mathematically. But every human being, even someone working in the field of quantum physics, lives and dies in position space, at least physically. So it is natural enough to find wave functions in the position representation particularly informative.

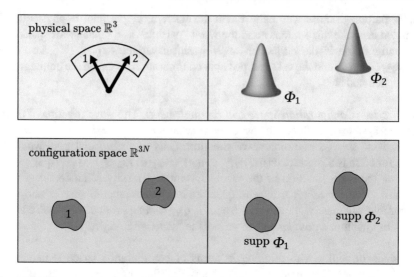

Fig. 1.1 Different macroscopic objects, here two physical pointers consisting of N atoms, where N has the order of magnitude of Avogadro's number, are represented by disjoint regions in the configuration space $\mathbb{R}^{3N}$ (*left*). The wave function of each pointer is concentrated in the corresponding area because of Born's interpretation of the wave function (*right*)

Those configurations that describe a pointer pointing to the left lie in a completely different region of configuration space than those that describe a pointer pointing to the right. And the two regions do not only differ by a few coordinates (or degrees of freedom), but by a huge number of them (order of magnitude 10^{24})—hence the term "macroscopically disjoint".

The wave functions of (macroscopic) bodies are functions on configuration space, and according to Born's rule—because we can see pointers or other macroscopic objects quite sharply—their wave functions are essentially only concentrated in the regions of configuration space which define the shapes, locations, and orientations of the bodies. Above we introduced the notion of support for the set of points where a function is *not* equal to zero. Thus, the wave functions that describe different pointer positions[12] will essentially have macroscopically disjoint supports (see Fig. 1.1). The configuration space of a single particle is the physical space $\mathbb{R}^3$. That is sufficient to understand the double-slit experiment, as discussed. But often in this context the question arises: If we do not close the slits but attach a measuring device designed to determine which slit the particle goes through, what happens

---

[12]Mathematically trained students will soon realise that if an initial wave function has compact support the time evolved wave function (solving Schrödinger's equation) will immediately have support equal to the whole of configuration space. But the function will be almost zero in most of configuration space. Hence all the following statements about disjoint supports and the associated picture can be taken with a grain of salt, i.e., they should be taken in the sense of physics and not pedantic mathematics.

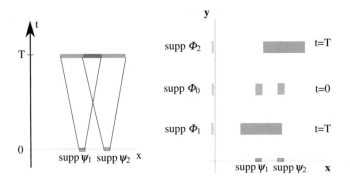

Fig. 1.2 *Left*: Superposition of two initially spatially separated wave functions of a system (**x**) which spread in time and whose supports overlap at time T. In the overlap region interference occurs. *Right*: In the configuration space of the apparatus (**y**) and the system, the coupling of a piece of apparatus with pointer wave functions Φ_0 (pointer null setting) and $\Phi_{1,2}$ (displays). The macroscopic wave functions have disjoint supports, which remain disjoint throughout the time evolution. The spreading of the system wave function parts does not lead to any overlap of the supports. Interference is not possible

then? The answer is that *decoherence* takes place and the interference pattern disappears. Simply put, the wave function of the particle becomes entangled with the wave function of the measurement device, and this entangled, high-dimensional wave function no longer interferes. Related to this is the phenomenon that the interference pattern becomes weaker or even disappears completely when very large molecules (say 10^{10} particles, sometimes referred to as Schrödinger cat states) are sent through the double slit.

The disappearance of interference can be explained by Fig. 1.2, which we will now discuss only briefly, because we will repeatedly take up the same reasoning in detail in the following chapters. A measuring device (symbolized by a pointer) is a macroscopic system and a measurement experiment involves an interaction between the system to be measured (e.g., the particle in the double-slit experiment) and the apparatus. The wave function of the total system (particle plus apparatus) evolves according to the Schrödinger equation (1.2) for the total system.

In the following, the reader will have to make some effort to make the connection between the abstract representation and the actual double-slit experiment. We consider a system (coordinates **x**, dimension m), whose wave function consists of a superposition of two wave functions $\psi_1(\mathbf{x})$ and $\psi_2(\mathbf{x})$ (think of the wave components, which start in each slit as Huygens' spherical waves), coupled to a piece of apparatus (coordinates **y**, dimension n), represented by the pointer null state $\Phi_0(\mathbf{y})$ and the two possible pointer positions $\Phi_1(\mathbf{y})$ and $\Phi_2(\mathbf{y})$. The coupling is expressed by an interaction potential occurring in the Schrödinger equation and it is assumed that the corresponding Schrödinger evolution[13] of the total wave function yields

$$\psi_i \Phi_0 \xrightarrow{t \longrightarrow T} \psi_i \Phi_i , \qquad i = 1, 2 , \tag{1.19}$$

where T represents the duration of the measurement. This means that the pointer positions correlate with the wave functions ψ_1 and ψ_2, respectively, the pointers "indicate either ψ_1 or ψ_2", e.g., whether the particle was registered at the upper or at the lower slit. If now the system has initially the wave function $\psi = \psi_1 + \psi_2$ (we ignore normalization constants), then due to linearity (see Remark 1.1) and using (1.19), we obtain

$$\psi \Phi_0 \xrightarrow{t \longrightarrow T} \psi_1 \Phi_1 + \psi_2 \Phi_2 . \tag{1.20}$$

Look at the corresponding Fig. 1.2 of the supports of the wave functions in configuration space! Actually, this contains everything we need to understand the phenomenon, but to really grasp that, we do need to think deeply about what the figure shows, because thinking in terms of configuration space is not easy. In any case, the conclusion is that the (apparatus + system) wave function parts have disjoint supports after the measurement and therefore pass each other in configuration space without overlapping. Therefore the interference terms (1.18) disappear and we have what is called a decoherent superposition. This does not explain why we always see only one of the two pointer positions (this is the measurement problem), but it does explain the disappearance of the interference pattern on the screen.[14]

Why does something similar happen, namely the suppression of interference, when large molecules are used in the double-slit experiment? Because a large molecule interacts more easily with its surroundings (air, electromagnetic waves, etc.) than a photon or an electron does. Loosely speaking, the environment "measures" the position of the molecule more effectively (like a piece of apparatus) than it does for a photon, for example. That means that interaction is ubiquitous, sometimes very invasive, sometimes less invasive. For macroscopic objects like pointers or cats, the interaction with the environment is very invasive, so interference of superpositions of macroscopic wave functions is practically impossible, because

---

[13]Starting with a "product wave function" $\Psi(\mathbf{q}) = \Psi(\mathbf{x}, \mathbf{y}) = \psi_i(\mathbf{x})\Phi_0(\mathbf{y})$ expresses the idea of the physical independence of system and apparatus before they interact. In the case of no interaction $V(\mathbf{x}, \mathbf{y}) \approx 0$, the Schrödinger equation (1.2) "factorises" into two independent equations, one for each factor of the product wave function. In that situation, the system and apparatus remain "physically independent". However, it is important to note that $V(\mathbf{x}, \mathbf{y}) \approx 0$ by itself does not imply physical independence because the wave function need not have product structure.

[14]Note that a superposition of wave functions does not necessarily mean that interference takes place. By the same token, if a superposition does not lead to interference, i.e., decoherence takes place, that does not mean that there is no superposition.

the wave parts entangle continuously with the wave function of the environment. Thus the effect of decoherence grows. We may now understand why we do not find such quantum interference effects in our macroscopic world of experiences, i.e., in classical physics, which naturally arose first as a description of nature. That is, we understand why we never have the experience of our interlocutor melting away like a character in an Edgar Allan Poe story.

Macroscopically separated wave components of a superposition run past each other (in configuration space). And the higher the dimensionality of the configuration space, the harder it is for them to "meet". If an experimenter were to set out to perform a truly macroscopic interference experiment (interference of the wave parts in Schrödinger's cat experiment), her task would be as complex as the following: simultaneously reverse the velocity of every gas molecule in a room. This analogy is apt, because it is about the precise control of roughly 10^{24} degrees of freedom. In the quantum mechanics of large systems, these are the phases in the macroscopic wave, in classical mechanics, the velocities of the particles. It can also be said that decoherence is an *irreversible process* in the thermodynamic sense. Since macroscopic wave function parts can no longer interfere and we always "see" only one of the macroscopic parts, one can introduce the famous collapse of the wave function, i.e., pretend that the branch of the wave function that we no longer see simply disappears, which sounds a bit like "putting one's head in the sand". We'll discuss this in more detail in Chap. 2.

1.6 The Classical Limit

This term so often used and heard doesn't actually make much sense. Limits are taken in mathematics for proving theorems and thus do not belong in the world of physics. What exactly does the "classical limit" mean then? The term is used as a shorthand description for physical situations that are more conveniently described by classical mechanics than by quantum mechanics. A bit more needs to be said because we seem to do immensely well with classical physics and it's usually much more difficult to see quantum mechanical effects. Nevertheless, it is generally accepted that quantum mechanics is more fundamental than classical mechanics, i.e., the latter should be included in the new theory as a kind of "limiting case".

It is sometimes said that the classical limit comes about by letting Planck's quantum of action $\hbar$ goes to zero. For example, in the limit $\hbar \to 0$, the commutator of the position and momentum operators becomes the Poisson bracket of position and momentum in classical mechanics. Why is that statement meaningless? Because Planck's constant is fixed at the value $2\pi\hbar = 6.62607015 \times 10^{-34}$ J s no matter what, so it cannot tend anywhere at all. So what is actually intended? There is indeed a lot to be said about a statement like this. For example, it may mean that in certain physical situations, where the evolution of the system is determined by a physical quantity with the dimension of an action (J s) and where the ratio of $\hbar$ to the value of this action is very small, the evolution of the system can be considered classically. And how small is small? So small that the wavy nature which in general determines the evolution of quantum objects can be ignored. That is the case when interference effects are suppressed.

Let us make a second comment. Classical physics deals with objects located in physical space, such as point particles in Newtonian physics. They move according to Newton's laws. That's classical physics. However, according to the doctrines of quantum theory (the Copenhagen interpretation, for example) localized objects (such as particles) do not exist. But if quantum theory is seen as fundamental (incorporating classical physics), how can such localised objects ever arise? That is mysterious. How much clearer the situation would be if there were already such localized objects in the more fundamental theory, whose motion in certain physical situations looked like the classical orbits of matter, e.g., in situations which could be characterized by the ratio of an action pertaining to the system and $\hbar$. And so we come back to the all-importance of *ontology*.

Unfortunately, one often hears or reads that the introduction of localized objects into quantum mechanics, such as the flashes of the GRW theory or the particles of Bohmian mechanics, would amount to a (stubborn) return to classical physics. So classical physics is taken as being synonymous with a physics of localized objects. This is unjustified, as the theories described in the following chapters will show.

1.6.1   Motion of Concentrated Wave Packets

A fundamental condition for the retrieval of classical mechanics in quantum mechanics is, according to Born's statistical hypothesis, to have a well localized wave function (a *wave packet*), because classical objects always seem to be well localized: a table may wobble because the floor is uneven, but it is never blurred or even in a superposition of table here and table there. If we move it, it moves along a Newtonian trajectory. Therefore, no matter which theory we prefer, we must eventually understand how well-localized wave functions move. When we discuss collapse models and Bohmian mechanics in the later chapters, by virtue of Born's statistical hypothesis the localized fundamental quantities such as the GRW flashes or particle positions in Bohmian mechanics will typically follow the center of the "well concentrated" wave packet.

But there are wave functions (actually most of them), which are not at all localised and therefore oppose classical physics in every respect, as for example (1.20). We discuss these in detail in Chap. 2. In order to cope with such wave functions, we must take the solutions to the measurement problem seriously, and if we do so, the remaining problems simply evaporate. This also includes what we have just shown in Sect. 1.3, namely that every wave packet spreads over time. How fast it spreads depends on parameters such as the mass, i.e., there are time scales (depending on the mass) over which the initially localized wave packet will remain well localized.

But spreading takes place as long as the wave packet evolves (more or less) freely, i.e., when the system can be considered as isolated. So does that mean that a classical description will not be possible after all? Well, wave packets are never completely isolated from the rest of the world. There are interactions with the surrounding air or with light and, as we said in the previous section, interactions can function like measurements. Here, once again, is the importance of the configuration space! That is, the environment "constantly checks where the particle is," and it always "collapses" to that part of the wave function in which the particle is located.

But isn't this all just idle talk? What exactly does it mean to say that the wave function collapses? And if the environment "measures" where the particle is, what or who "measures" the environment? So we end up once again with the situation of Chap. 2 and we must understand the possible solutions of the measurement problem. There is simply no way around that.

But for now we do not care how long an initially concentrated wave function remains concentrated, we just follow it as long as it stays concentrated enough, where "enough" can be quantified to some extent by the following. We model the evolution of the wave function in an external potential V, i.e., the wave function $\psi(\mathbf{x})$ evolves according to the Schrödinger equation (1.2):

$$i\hbar \frac{\partial \psi}{\partial t}(\mathbf{x}, t) = -\frac{\hbar^2}{2m} \Delta \psi(\mathbf{x}, t) + V(\mathbf{x})\psi(\mathbf{x}, t) =: H\psi, \tag{1.21}$$

where the symbol H denotes what is called the *Hamiltonian*. Born's statistical hypothesis states that the expected value of the position evolves according to[15]

$$\langle \mathbf{X} \rangle (t) := \int \mathbf{x} |\psi(\mathbf{x}, t)|^2 \, \mathrm{d}^3 x. \tag{1.22}$$

Taking the derivative with respect to time and using (1.4), we get

$$\frac{\mathrm{d}}{\mathrm{d}t} \langle \mathbf{X} \rangle (t) = \int \mathbf{x} \frac{\partial}{\partial t} |\psi(\mathbf{x}, t)|^2 \, \mathrm{d}^3 x$$

$$= -\int \mathbf{x} \nabla \cdot \mathbf{j}(\mathbf{x}, t) \mathrm{d}^3 x$$

$$= \int \mathbf{j}(\mathbf{x}, t) \, \mathrm{d}^3 x, \tag{1.23}$$

after integrating by parts and setting the boundary terms to zero. In classical physics, i.e., Newtonian physics, acceleration plays the key role in the law of motion. Therefore we are interested in $\mathrm{d}^2 \langle \mathbf{X} \rangle / \mathrm{d}t^2$, i.e.,

$$\frac{\mathrm{d}^2}{\mathrm{d}t^2} \langle \mathbf{X} \rangle (t) = \int \frac{\partial}{\partial t} \mathbf{j}(\mathbf{x}, t) \, \mathrm{d}^3 x.$$

---

[15]The symbol $\langle \cdot \rangle$ is a common way to denote a mean or average.

Recalling the quantum flux equation (1.5), we have

$$\frac{\partial}{\partial t}\mathbf{j} = -\frac{i\hbar}{2m}\frac{\partial}{\partial t}\left(\psi^*\nabla\psi - \psi\nabla\psi^*\right).$$

Observing that ψ^* satisfies the complex-conjugated Schrödinger equation [take the conjugate complex of (1.21)], we obtain

$$\frac{\partial}{\partial t}\mathbf{j} = \frac{1}{2m}\left[(H\psi^*)\nabla\psi - \psi^*\nabla(H\psi) + (H\psi)\nabla\psi^* - \psi\nabla(H\psi^*)\right].$$

Observing further that, for any ψ, φ and using partial integration and the self-adjointness of H,

$$\int \psi^* H\varphi \, d^3x = \int (H\psi^*)\varphi \, d^3x,$$

we get

$$
\begin{aligned}
\frac{d^2}{dt^2}\langle\mathbf{X}\rangle(t) &= \frac{1}{2m}\int\left[\psi^* H\nabla\psi - \psi^*\nabla(H\psi) + \psi H\nabla\psi^* - \psi\nabla(H\psi^*)\right]d^3x \\
&\overset{(1.21)}{=} \frac{1}{2m}\int\left[\psi^* V\nabla\psi - \psi^*\nabla(V\psi) + \psi V\nabla\psi^* - \psi\nabla(V\psi^*)\right]d^3x \\
&= \frac{1}{2m}\int\left[-(\nabla V)\psi^*\psi - (\nabla V)\psi\psi^*\right]d^3x \\
&= \frac{1}{m}\langle -\nabla V(\mathbf{X})\rangle(t).
\end{aligned}
$$

Amazing, isn't it? We get the Newtonian equations "on average". This is a version of the famous Ehrenfest theorem:

$$m\langle\ddot{\mathbf{X}}\rangle(t) = \langle -\nabla V(\mathbf{X})\rangle(t). \tag{1.24}$$

We would get the classical limit and with it the identification of the parameter m as the Newtonian mass if we also had

$$\langle -\nabla V(\mathbf{X})\rangle(t) \approx -\nabla V(\langle\mathbf{X}\rangle(t)), \tag{1.25}$$

because then we would get the Newtonian equation of motion for $\langle\mathbf{X}\rangle$. This is where $\psi(\mathbf{x}, t)$ being "concentrated enough" comes into play: recalling the definition (1.22) of the expected value, we see that the fluctuation around the average value is small, so the expected value of the function of the random variable is given approximately by the function of the expected value.

As a final remark, or better, as a warning, we said earlier that the spreading of the wave function is countered by the interaction with the environment, which ensures that the wave packet remains well localized.[16] But why, we may wonder, is the Schrödinger equation still applicable, as we have assumed in the above argument? In fact in such situations, where it is the environment which produces the wave packet, it is no longer valid for the wave packet because it is not a closed system. This means that we still need to justify its approximate validity if we are to ensure that the above computation is well founded. We shall not do that here. We refer instead to *Bohmian Mechanics. The Physics and Mathematics of Quantum Theory.*[17]

1.7   Spin and the Stern–Gerlach Experiment

Another innovation of quantum mechanics is "spin". It is tempting, given the name, to think of a continuously rotating object, but that is not appropriate. Nevertheless, it is easy to say what spin is mathematically. What is not as easy is to explain why, for example, an electron "must have" spin. The "nonclassical bivalence" [as named by Wolfgang Pauli (1900–1958)] of the spin $-1/2$ electron is something really new, something fundamentally quantum mechanical. In the Stern–Gerlach experiment, a silver atom[18] passing through a Stern–Gerlach magnet (which produces a strongly inhomogeneous magnetic field) is deflected like a magnetic dipole which adjusts itself to lie parallel (spin $+1/2$) or antiparallel (spin $-1/2$) to the field gradient (see Fig. 10.1 for an experiment with the Stern–Gerlach setup). A Stern–Gerlach system which is set up "in the z-direction" prepares "spin $+1/2$ particles" or "spin $-1/2$ particles" in the z-direction. Now send a z-spin $+1/2$ particle through a Stern–Gerlach apparatus which is oriented both (1) orthogonal to the z-direction and (2) orthogonal to the flight direction of the particle. Let us say that it is oriented in the y-direction. Then with probability $1/2$ the particle will end up as a y-spin $+1/2$ or a y-spin $-1/2$ particle. If we choose any other direction, only the probabilities for the splits are influenced, but the bivalence remains. This is an experimental fact.

---

[16]A warning within a warning: the interaction with the environment actually produces a measurement-like situation, as discussed in Chap. 2, i.e., we end up with a superposition of wave packets which are more or less spatially separated. So Chap. 2 is relevant for this too.

[17]D. Dürr and S. Teufel, *Bohmian Mechanics. The Physics and Mathematics of Quantum Theory.* Springer, 2009.

[18]Actually one would like to do the experiment with electrons, but silver atoms have the advantage of being electrically neutral while possessing a net magnetic moment, due to a single outer electron. Therefore, it is as good as an electron as far as spin is concerned. The fact that it is electrically neutral is also good, since the Lorentz force on charged particles does not then contribute.

The wave function of the particle must therefore observe this splitting and the simplest way to achieve this is to assign two degrees of freedom to the ψ-function itself. ψ thus becomes a *spinor wave function*:

$$\psi : \mathbb{R}^3 \to \mathbb{C}^2, \quad \psi(\mathbf{x}) = \begin{pmatrix} \psi_1(\mathbf{x}) \\ \psi_2(\mathbf{x}) \end{pmatrix}, \tag{1.26}$$

with normalisation

$$\int (\psi, \psi)(\mathbf{x}) \, d^3 x = \int \begin{pmatrix} \psi_1^*(\mathbf{x}) \\ \psi_2^*(\mathbf{x}) \end{pmatrix} \cdot \begin{pmatrix} \psi_1(\mathbf{x}) \\ \psi_2(\mathbf{x}) \end{pmatrix} d^3 x = \int \left(|\psi_1(\mathbf{x})|^2 + |\psi_2(\mathbf{x})|^2 \right) d^3 x = 1 .$$

A spinor need not be two-dimensional like here. Other dimensions are possible, for example four, for the relativistic Dirac equation [Paul Dirac (1902–1984), see Sect. 11.1.2] for an electron.

The spin degrees of freedom couple to the magnetic field, as the Stern–Gerlach experiment suggests. This is how. The Schrödinger equation for the wave function will be replaced by the so-called *Pauli equation*, where the "potential" V becomes a Hermitian matrix, and such a matrix can always be written in the form

$$V(\mathbf{x}) E_2 + \mathbf{B}(\mathbf{x}) \cdot \boldsymbol{\sigma}, \quad V(\mathbf{x}) \in \mathbb{R}, \quad \mathbf{B}(\mathbf{x}) \in \mathbb{R}^3, \quad \boldsymbol{\sigma} = (\sigma_x, \sigma_y, \sigma_z),$$

where we define the Pauli matrices

$$\sigma_x = \begin{pmatrix} 0 & 1 \\ 1 & 0 \end{pmatrix}, \quad \sigma_y = \begin{pmatrix} 0 & -i \\ i & 0 \end{pmatrix}, \quad \sigma_z = \begin{pmatrix} 1 & 0 \\ 0 & -1 \end{pmatrix}. \tag{1.27}$$

The Pauli equation reads

$$i\hbar \partial_t \psi = \left[\frac{1}{2m} \left(-i\hbar \nabla - e\mathbf{A} \right)^2 + eV \right] \psi - \mu \boldsymbol{\sigma} \cdot \mathbf{B} \psi, \tag{1.28}$$

where we use the following notation:

- $\psi(\mathbf{x}) = \begin{pmatrix} \psi_1(\mathbf{x}) \\ \psi_2(\mathbf{x}) \end{pmatrix}$ is the spinor wave function.
- e a parameter to be identified with the charge of the particle.
- μ is a "coupling constant", called the gyromagnetic factor, which acts like the strength of the magnetic moment.
- $\mathbf{A}$ is the vector potential.
- V is the electrostatic potential.
- $\mathbf{B} = \nabla \times \mathbf{A}$ is the magnetic field.

In the normalisation of the spinor wave function [below (1.26)], we have used the *scalar product in spin space*:

$$(\psi, \psi)(\mathbf{x}) := \left(\psi_1^*(\mathbf{x}), \psi_2^*(\mathbf{x})\right) \begin{pmatrix} \psi_1(\mathbf{x}) \\ \psi_2(\mathbf{x}) \end{pmatrix} = \sum_{i=1,2} \psi_i^*(\mathbf{x})\psi_i(\mathbf{x}) . \tag{1.29}$$

This enters Born's statistical distribution for the position of the particle which, for the Pauli equation, becomes $\varrho^\psi = (\psi, \psi)$. To see this we show that $\varrho^\psi = (\psi, \psi)$ satisfies the continuity equation (1.4) with

$$\mathbf{j}^\psi = \frac{\hbar}{2im}\left[(\psi, \nabla\psi) - (\nabla\psi, \psi)\right] = \frac{\hbar}{m}\mathrm{Im}\,(\psi, \nabla\psi) . \tag{1.30}$$

The required computation is very much analogous to the one leading to (1.4), but replacing the real function V by a Hermitian one, which does not affect the argument.

Remark 1.5 (The Pauli Flux)   What we just said is only for starters and we need to add the caveat that there is a debate about what should be considered the right flux, or what should be the physical flux for the Pauli equation. The general idea is that it is not given by (1.30). There is nothing wrong with the computation suggested above, but we must note that the continuity equation contains the divergence of $\mathbf{j}^\psi$, with the consequence that the flux vector is not uniquely fixed by the continuity equation. Any divergence-free vector field can be added to the flux vector, for example, the curl of a vector field $\mathbf{a}(\mathbf{x})$, i.e., $\nabla \times \mathbf{a}(\mathbf{x})$, since by simple vector analysis $\nabla \cdot \nabla \times \mathbf{a}(\mathbf{x}) = \nabla \times \nabla \cdot \mathbf{a}(\mathbf{x}) = 0$. Is that a disaster? No, not really. We can just note this and stay cool. If we wish to mess around with the Schrödinger flux then there are the physical spacetime symmetries which must be respected by additional terms, and this renders the Schrödinger flux pretty much unique.

But in the Pauli situation there does exist a natural additional flux term which is well behaved in all respects and which is sometimes referred to as the Gordon term [Walter Gordon (1883–1939), known from the Klein–Gordon equation] or the *spin flux*. For a one-particle Pauli equation, this reads

$$\frac{\hbar}{2m} \nabla \times (\psi, \boldsymbol{\sigma}\psi) .$$

When added to the so-called *convective flux* (1.30), the total flux is

$$\mathbf{J}^\psi{}_{\mathrm{Pauli}} = \frac{\hbar}{m}\left[\mathrm{Im}(\psi, \nabla\psi) + \frac{1}{2}\nabla \times (\psi, \boldsymbol{\sigma}\psi)\right] . \tag{1.31}$$

There are many arguments in the physics literature to justify taking this as the correct Pauli flux. A straightforward and simple one comes with the derivation of the Pauli equation from the relativistic Dirac equation. In particular, the Pauli equation results from the relativistic Dirac equation in a "non-relativistic limit".[19] The Dirac equation is the relativistic wave equation for an electron, and when Dirac invented it, the Schrödinger wave was replaced by a spinor wave function which naturally had four components, and not two as in the Pauli equation. The Pauli equation arises in this view as a "non-relativistic limit" of the Dirac equation, and when we take the steps required to extract the Pauli equation and consider how this affects the Dirac flux (11.5), which itself satisfies the continuity equation (11.6) for the Dirac equation, the Pauli flux $\mathbf{J}^{\psi}{}_{\text{Pauli}}$ emerges as the "non-relativistic limit" of the Dirac flux.[20]

We mentioned in Sect. 1.4 that the quantum flux provides the statistics of arrival times. In this respect it is interesting to ask whether the arrival times of spin-1/2 particles are determined by the correct Pauli flux. Quantum measurements of first arrival times would provide helpful information, but they are still lacking.[21]

1.7.1   The Pauli Equation and the Stern–Gerlach Experiment

The reasoning behind the appearance of a two-component spinor was based on the experimental fact that the wave function splits in a Stern–Gerlach experiment. This fact will be used in Chap. 10 in the presentation of the famous EPRB argument by Einstein, Podolsky, and Rosen, but in the spin version proposed by Bohm. Therefore we wish to understand at least heuristically how the splitting comes about by examining the Pauli equation.

As noted above, the Stern–Gerlach experiment is done with neutral atoms. Therefore in what follows we shall consider a neutral particle and set $e = 0$ in (1.28). We consider a Stern–Gerlach setup in which the particle crosses the magnetic field in the y-direction, while the magnetic poles are directed in z-direction. We idealize the experiment as taking place in the (y, z)-plane. The magnetic field is then $\mathbf{B} = (0, B_y(y, z), B_z(y, z))$ with div $\mathbf{B} = 0$. Since the field is inhomogeneous in the z-direction, this implies that there is also an inhomogeneous field in the y-direction, but we shall ignore this effect. Furthermore, we shall see that the deflection which we wish to argue for depends on the derivative $\partial B_z / \partial z$. So to a first approximation, we can look at the first order term in the Taylor expansion of $B_z(y, z)$ around the median value of z (midway between the two poles), which we can set equal to zero. For simplicity, we can also set $B_z(y, 0) = 0$. So we end up with the approximation

$$\mathbf{B} \approx \left(0, B_y(y), b(y)z\right),$$

[19]We need to spell out carefully what is meant by this, but superficially we may consider that in a particular physical situation the speed of the electron is much less than the speed of light.

[20]See, e.g., M. Nowakowski, The quantum mechanical current of the Pauli equation. Am. J. Phys. **67**, 916–919 (1999), and for an overview, W.B. Hodge, S.V. Migirditch, W.C. Kerr, Electron spin and probability current density in quantum mechanics. Am. J. Phys. **82**, 681–690 (2014).

[21]For more on this, see S. Das and D. Dürr, Arrival time distributions of spin-1/2 particles. Sci. Rep. **9**, 2242 (2019), and arXiv:1802.07141.

where we think of $b(y)$ as an indicator function which has a constant value b in the region of the magnetic field and vanishes outside.

Bearing in mind that we are considering a planar situation, let the initial wave function of the spin particle be

$$\Phi_0(\mathbf{x}) = \varphi_0(z)\psi_0(y)\left[\alpha\begin{pmatrix}1\\0\end{pmatrix} + \beta\begin{pmatrix}0\\1\end{pmatrix}\right], \tag{1.32}$$

where $\begin{pmatrix}1\\0\end{pmatrix}$ and $\begin{pmatrix}0\\1\end{pmatrix}$ are eigenvectors of σ_z with eigenvalues[22] 1 and -1, respectively. (For ease of notation, we have assumed the same position dependence in the wave function for both spinors. The following argument does not hinge on this.)

In the approximation of the magnetic field above, the y-motion of the wave packet essentially decouples from the z-motion inside the region where the magnetic field acts, so we obtain two Pauli equations, one depending only on the y-variable and one depending only on the z-variable, with the latter containing the interaction term $\mu\sigma_z bz$. Moreover, we assume that the wave packet moves fast in the y-direction and is centred around the median. It will spend a short time τ in the magnetic field. That is all we use from the y part of the problem. We focus now only on the z-motion of the wave packet in the magnetic field. In doing so we also ignore the spreading of the wave due to the Laplace term in the Pauli equation (1.28). Since the equation is linear we need only consider $\Phi^{(1)}(z) = \varphi(z)\begin{pmatrix}1\\0\end{pmatrix}$ and $\Phi^{(2)}(z) = \varphi(z)\begin{pmatrix}0\\1\end{pmatrix}$. We represent the initial wave packet in its Fourier modes

$$\varphi_0(z) = \int e^{ikz} f(k)dk\,, \quad f(k) \text{ concentrated around } k_0 = 0\,, \tag{1.33}$$

which means in particular that there is no initial momentum in the z-direction. Thus the problem is reduced to the solutions of the approximate Pauli equation, viz., (1.28) without the Laplace term, $e = 0$ and with $\mu\boldsymbol{\sigma}\cdot\mathbf{B}$ replaced by $\mu bz\sigma_z$:

$$i\hbar\frac{\partial\Phi}{\partial t}(z,t) \approx -\mu bz\,\sigma_z\Phi(z,t)\,. \tag{1.34}$$

This means the spinor parts satisfy

$$i\hbar\frac{\partial\Phi^{(n)}}{\partial t}(z,t) = (-1)^n\mu bz\,\Phi^{(n)}(z,t)\,, \quad n = 1,2\,. \tag{1.35}$$

After a time of flight $t > \tau$, that is, after the wave packet has left the magnetic field, we have

$$\Phi^{(n)}(z,t) = e^{-i(-1)^n\frac{\mu b\tau}{\hbar}z}\Phi_0^{(n)}(z)\,. \tag{1.36}$$

---

[22]We talked above about spin-1/2. For our purposes, this particular value is merely an unimportant convention and has been absorbed in the gyromagnetic factor μ.

The z-wave packets (1.33) now evolve freely again with wave numbers

$$\tilde{k} = k - (-1)^n \frac{\mu b \tau}{\hbar} ,$$

and since they are solutions of the free wave equations, the frequencies are given by the free dispersion relation $\omega(\tilde{k}) = \hbar^2 \tilde{k}^2 / 2m$. Noting that $k_0 = 0$, the group velocity is

$$\left. \frac{\partial \omega(\tilde{k})}{\partial \tilde{k}} \right|_{k_0=0} = -(-1)^n \frac{\hbar \mu b \tau}{m} .$$

Therefore, the wave packet $\Phi^{(1)}$ moves along the positive z-direction, i.e., in the direction of the gradient of $\mathbf{B}$ (by convention we call this *spin up*) and $\Phi^{(2)}$ moves along the negative z-direction (by convention we call this *spin down*).

We find the spatial separation of the initial wave packet (1.32) into two parts along the z-axis, whence the initial wave packet evolves into a superposition of two separating wave packets, where the separation continues until new interactions occur. That is, ignoring the y degrees of freedom,

$$\Phi(z) = \alpha \varphi^{(1)}(z) \begin{pmatrix} 1 \\ 0 \end{pmatrix} + \beta \varphi^{(2)}(z) \begin{pmatrix} 0 \\ 1 \end{pmatrix} , \tag{1.37}$$

where $\varphi^{(1)}$ is concentrated above and $\varphi^{(2)}$ is concentrated below the initial flight axis. According to Born's rule (1.2), and taking $\varphi^{(1)}$ and $\varphi^{(2)}$ to be normalised, we conclude that with probability $|\alpha|^2$ the particle will be above, i.e., in the support of $\varphi^{(1)}$—then we say that the particle has spin up—and with probability $|\beta|^2$ it will be below, i.e., in the support of $\varphi^{(2)}$—then we say that the particle has spin down.

For a general normalized spinor wave function

$$\begin{pmatrix} \psi_1(\mathbf{x}) \\ \psi_2(\mathbf{x}) \end{pmatrix} ,$$

we obtain analogously z-spin up with probability

$$\int \left| \begin{pmatrix} \psi_1(\mathbf{x}) \\ \psi_2(\mathbf{x}) \end{pmatrix} \cdot \begin{pmatrix} 1 \\ 0 \end{pmatrix} \right|^2 d^3x = \int |\psi_1|^2(\mathbf{x}) \, d^3x ,$$

and z-spin down with probability

$$\int \left| \begin{pmatrix} \psi_1(\mathbf{x}) \\ \psi_2(\mathbf{x}) \end{pmatrix} \cdot \begin{pmatrix} 0 \\ 1 \end{pmatrix} \right|^2 d^3x = \int |\psi_2|^2(\mathbf{x}) \, d^3x .$$

This notation, where on the left we recognize the projections onto the eigenvectors of σ_z, can be straightforwardly generalized to arbitrary orientations of the Stern–Gerlach magnets.

If the Stern–Gerlach magnet is oriented along $\mathbf{a}$ then the eigenvectors of σ_z will be replaced by the eigenvectors of $\mathbf{a} \cdot \boldsymbol{\sigma}$, usually denoted by $|\uparrow_{\mathbf{a}}\rangle$ and $|\downarrow_{\mathbf{a}}\rangle$). The probabilities of $\mathbf{a}$-spin up and $\mathbf{a}$-spin down will then be computed by projections onto the corresponding spin components, which we shall denote by $|\langle\uparrow_{\mathbf{a}}|\psi\rangle|^2$ and $|\langle\downarrow_{\mathbf{a}}|\psi\rangle|^2$, respectively.

1.8 Why "Spinors"?

Why is the wave function (1.26) said to be spinor-valued and not simply $\mathbb{C}^2$-valued? Because it is not just two-valuedness which plays a role, but also the *transformation property* under symmetry transformations. The Schrödinger and Pauli wave equations must respect the symmetries of Galilean spacetime. Among these are spatial rotations, and we need to specify how wave functions transform under spatial rotations, so that the equations for them remain invariant. But before we come to that, let us clarify what is actually meant when we speak of the symmetries of Galilean spacetime. For example, speaking about rotations in a pictorial manner, we can say that our physical space has no preferred direction, so that the physical law for the motion of "stuff" must not prefer one spatial direction over another. But how are rotations described mathematically, since mathematics is the language in which the book of nature is written? For that we need to recall some linear algebra.

Three-dimensional physical space is usually represented by the vector space $\mathbb{R}^3$. And when we think of rotations, we naturally think of rotating the direction of a vector through an angle. However, the ingenious discovery of Hermann Grassmann (1809–1877), the father of modern linear algebra, was that the fundamental meaning of a vector must be understood without any reference to angles. The latter are brought in by introducing an extra structure, which one imposes on the vector space. That extra structure is called a *scalar product*. The scalar product gives meaning to angles and rotations. In short, in Grassmann's abstract theory, rotations are linear transformations on the vector space which leave the Euclidean scalar product invariant. These linear transformations are represented by matrices in $\mathbb{R}^{3,3}$, which are orthogonal and special in the sense of having determinant 1. They form a group called $SO(3)$.

Next recall the notion of a vector field $\mathbf{F} : \mathbb{R}^3 \to \mathbb{R}^3$, as used in classical physics. At each point $\mathbf{x} \in \mathbb{R}^3$ there sits a vector

$$\mathbf{F}(\mathbf{x}) = \begin{pmatrix} F_1(\mathbf{x}) \\ F_2(\mathbf{x}) \\ F_3(\mathbf{x}) \end{pmatrix}$$

which carries a geometrical meaning, namely a *direction* and a *length*. This geometrical meaning manifests itself by the transformation property of the vector field. In short, rotating the coordinate system, the coordinate vectors at every point must rotate as well. Consider for example the Newtonian equations of motion of a particle in a force field $\mathbf{F}$:

$$m\ddot{\mathbf{x}} = \mathbf{F}(\mathbf{x}) .$$

If it is to be taken as fundamental, this equation should be invariant under rotations. Hence, if $\mathbf{x}(t)$, $t \in \mathbb{R}$ is a solution of Newton's equation, then $\tilde{\mathbf{x}}(t) := R\mathbf{x}(t)$ must be as well, where $R \in SO(3)$ is a three-dimensional rotation matrix. We can understand this in the sense of a *passive* transformation, in which case $\tilde{\mathbf{x}}(t)$ is physically *the same* solution trajectory, only represented in a coordinate system which has been rotated by R. The Newtonian equation of motion becomes

$$m\frac{\mathrm{d}^2}{\mathrm{d}t^2}\tilde{\mathbf{x}}(t) = mR\ddot{\mathbf{x}} = R\mathbf{F}(\mathbf{x}) = R\mathbf{F}(R^{-1}\tilde{\mathbf{x}}) \overset{!}{=} \tilde{\mathbf{F}}(\tilde{\mathbf{x}}) . \qquad (1.38)$$

Hence, if the spatial coordinates are transformed by the rotation matrix R, i.e., $\mathbf{x} \to \tilde{\mathbf{x}} = R\mathbf{x}$, then $\mathbf{F}(\mathbf{x})$ must also be transformed by R at each point, i.e., as $\mathbf{F}(\cdot) \to \tilde{\mathbf{F}}(\cdot) = R\mathbf{F}(R^{-1}\cdot)$. Such a field $\mathbf{F}$ is called a vector field.

The rotation matrices R are representations of the rotation group on the vector space $\mathbb{R}^3$. To make the connection with $\mathbb{C}^2$-valued spinors, we first build a small bridge. We recall that the plane $\mathbb{R}^2$ is isomorphic to the vector space $\mathbb{C}$. A rotation in the plane through an angle α is represented by a special (i.e., determinant 1) orthogonal 2×2 matrix, which means to say that it is in $\in SO(2)$, while in $\mathbb{C}$ the same rotation will be represented by multiplication by $e^{i\alpha}$. We thus make the following observation: instead of a two-dimensional representation of the rotations in the plane, we may as well choose a one-dimensional representation in $\mathbb{C}$ given by $e^{i\alpha}$, $\alpha \in [0, 2\pi)$. This representation is called the unitary one-dimensional group $U(1)$.

Spinors do something similar, but we need to dig a bit deeper to find a connection between $\mathbb{R}^3$ and $\mathbb{C}^2$. This connection is closely related to the field of quaternions and its isomorphism with $\mathbb{R}^4$, but it is not absolutely necessary to know about that if we use a bit of imagination. It is easy to see that vectors in $\mathbb{R}^3$ can be represented in the following way by Hermitian 2×2 matrices: $\mathbf{x} = (x, y, z) \in \mathbb{R}^3$ can be represented by

$$\mathbf{x} \cong x\sigma_1 + y\sigma_2 + z\sigma_3 = \begin{pmatrix} z & x - iy \\ x + iy & -z \end{pmatrix} ,$$

where we have used the Pauli matrices $\sigma_1, \sigma_2, \sigma_3$ as a basis. The representation is obviously one to one. Now the question is: How can we represent rotations in this case? Let us look at a rotation of the vectors through an angle α about the z-axis:

$$\begin{pmatrix} x' \\ y' \\ z' \end{pmatrix} = \begin{pmatrix} \cos\alpha & -\sin\alpha & 0 \\ \sin\alpha & \cos\alpha & 0 \\ 0 & 0 & 1 \end{pmatrix} \begin{pmatrix} x \\ y \\ z \end{pmatrix}.$$

To have that rotation act on the matrix representation of the vector $\mathbf{x}$, we recall that transformations of matrices always act in a two-sided way, with one transformation matrix multiplying from the left and the transpose or Hermitian conjugate of the transformation matrix multiplying from the right, in the case of orthogonal or unitary transformations, respectively. Since the transformation of the vector representation must preserve the Hermitian character, it must be a unitary 2×2 matrix U depending on α which effects the transformation. Unitary means that $U^+ = U^{-1}$, whence the determinant is ± 1. With the choice $\det U = 1$, the set of such matrices forms a special group under multiplication and is denoted by $SU(2)$. Later we shall say more about the connection between $SO(3)$ and $SU(2)$. Note as an aside that the Euclidean length of a vector is not changed under rotation. Here the length squared of $\mathbf{x}$ is given by

$$\|\mathbf{x}\|^2 = \left| \det \begin{pmatrix} z & x - iy \\ x + iy & -z \end{pmatrix} \right| = x^2 + y^2 + z^2,$$

so it is immediate from the rules about determinants that the matrix transformation does not change the length.

Following the above argument the $\mathbf{x}$-matrix will be multiplied by U from the left and by U^+ from the right, so that in each of the matrices only half of the angle must appear. It is easy to guess how the rotation by α around the z-axis should be represented, viz.,

$$U(\alpha) = \begin{pmatrix} e^{i\alpha/2} & 0 \\ 0 & e^{-i\alpha/2} \end{pmatrix}, \tag{1.39}$$

and we do indeed obtain the transformed vector (1.8)

$$\begin{pmatrix} z' & x' - iy' \\ x' + iy' & -z' \end{pmatrix} = \begin{pmatrix} e^{-i\alpha/2} & 0 \\ 0 & e^{i\alpha/2} \end{pmatrix} \begin{pmatrix} z & x - iy \\ x + iy & -z \end{pmatrix} \begin{pmatrix} e^{i\alpha/2} & 0 \\ 0 & e^{-i\alpha/2} \end{pmatrix}. \tag{1.40}$$

The reader is invited to check the details of the calculation.

Remark 1.6 (Two Possibilities) Note that (1.40) shows that $-U$ does the very same job, i.e., both $\pm U$ represent the same rotation. This double representation will be discussed again later.

But our aim is to get the transformation behaviour of the spinor! To do this we invoke a helpful mnemonic, namely that a spinor is something like a square root of a vector, in the sense that the "product" of two spinors yields a vector. A nice way to understand this is by rewriting the 2×2 Hermitian matrix representing our vector in terms of *dyadic products* of its eigenvectors. Recall that a Hermitian matrix has real eigenvalues, say λ_1 and λ_2, and two corresponding orthogonal eigenvectors, say s_1 and s_2. We normalize these. For $s = \begin{pmatrix} a \\ b \end{pmatrix} \in \mathbb{C}^2$, let $s^+ := (a^*\, b^*)$ be the transposed complex conjugated vector. Then,

$$\begin{pmatrix} z & x - iy \\ x + iy & -z \end{pmatrix} = \lambda_1 s_1 s_1^+ + \lambda_2 s_2 s_2^+ , \qquad (1.41)$$

as can be seen by working out the eigenvalues and eigenvectors and checking that everything comes out just right. Here is an example (where the eigenvalues have been absorbed into the eigenvectors):

$$\begin{pmatrix} 0 & -iy \\ iy & 0 \end{pmatrix} = \begin{pmatrix} -i\sqrt{y/2} \\ \sqrt{y/2} \end{pmatrix} \left(i\sqrt{y/2}, \sqrt{y/2} \right) - \begin{pmatrix} i\sqrt{y/2} \\ \sqrt{y/2} \end{pmatrix} \left(-i\sqrt{y/2}, \sqrt{y/2} \right) .$$

The matrix product of a vector $s = \begin{pmatrix} a \\ b \end{pmatrix} \in \mathbb{C}^2$ and the dual vector t^+ of another vector $t = \begin{pmatrix} c \\ d \end{pmatrix} \in \mathbb{C}^2$, viz.,

$$st^+ = \begin{pmatrix} a \\ b \end{pmatrix} (c^*\, d^*) = \begin{pmatrix} ac^* & ad^* \\ bc^* & bd^* \end{pmatrix} ,$$

is called the *dyadic product*. So we see that a three-dimensional Euclidean vector can be represented as a Hermitian matrix, and the Hermitian matrix can be decomposed into dyadic products of two-dimensional complex vectors s_1 and s_2. Now, s_1 and s_2 are spinors!

From this representation we also see why left–right multiplication by the rotation matrix in (1.40) makes sense. In particular, the spinor s transforms only by multiplying by U from the left. Then s^+ automatically transforms by multiplying by U^+ from the right.

We now apply these insights to the spinor wave function to formulate the transformation law. At each point $\mathbf{x}$, the wave function $\psi(\mathbf{x})$ has a geometrical meaning, a "direction" as it were, and we need to say how $\psi(\mathbf{x})$ transforms when we rotate the coordinate system. The argument goes like this. Let $R(\alpha, \mathbf{n})$ be the three-dimensional rotation matrix for the rotation through the angle α about the rotation axis $\mathbf{n} \in \mathbb{R}^3$, where $\|\mathbf{n}\| = 1$. The reader should check that the corresponding complex rotation matrix generalising (1.39) which acts on the spinor reads

$$U(\alpha, \mathbf{n}) := \exp\left(-i\frac{\alpha}{2}\mathbf{n} \cdot \boldsymbol{\sigma}\right), \tag{1.42}$$

where $\mathbf{n} \cdot \boldsymbol{\sigma} = n_1\sigma_1 + n_2\sigma_2 + n_3\sigma_3$. For a coordinate transformation

$$\mathbf{x} \to \tilde{\mathbf{x}} = R(\alpha, \mathbf{n})\mathbf{x},$$

the spinor wave function transforms according to [compare with (1.38)]

$$\psi(\mathbf{x}) \to \tilde{\psi}(\tilde{\mathbf{x}}) = \exp\left(-i\frac{\alpha}{2}\mathbf{n} \cdot \boldsymbol{\sigma}\right)\psi(R^{-1}\tilde{\mathbf{x}}). \tag{1.43}$$

Such a ψ is called a spinor field, or a spinor wave function in the present case. Note that the $\mathbb{C}^1$-valued wave function (1.1) allows only a trivial representation of the rotations in $\mathbb{R}^3$, which means that it transforms as a *scalar*: $\tilde{\psi}(\tilde{\mathbf{x}}) = \psi(R^{-1}\tilde{\mathbf{x}}) = \psi(\mathbf{x})$.

Note that, in the Pauli flux vector $\mathbf{J}^\psi{}_{\text{Pauli}}(\mathbf{x})$ given in (1.31), the spinor wave function also enters in a bilinear way, so once again it arises as something like a square root of a vector, but not quite in the way we discussed above. It is an enlightening exercise to actually show how the transformation property (1.43) of the spinor gives rise to the correct rotation of the Pauli flux vector $\mathbf{J}^\psi{}_{\text{Pauli}}(\mathbf{x})$.

It is worth saying more about all this. An interesting property of the spinor representation (1.42) is that a rotation through the angle 2π about any axis amounts simply to multiplication by -1. The simplest way to check this is to look at the rotation about the z-axis for our choice of the Pauli matrices. Since

$$\sigma_z = \begin{pmatrix} 1 & 0 \\ 0 & -1 \end{pmatrix}$$

is diagonal, we have

$$\exp\left(-i\frac{2\pi}{2}\sigma_z\right) = \begin{pmatrix} e^{-i\pi} & 0 \\ 0 & e^{i\pi} \end{pmatrix} = \begin{pmatrix} -1 & 0 \\ 0 & -1 \end{pmatrix}.$$

This means that, under a full rotation through $360°$, the spinor wave function changes sign. This has no empirical consequences because the absolute square does not change. Only after *two* full rotations does the symmetry transformation (1.42) once again represent the identity. But isn't it strange if, after a full rotation, an object does not return to its initial state? Not so strange, if one remembers the mnemonic that a spinor is a bit like the square root of a vector.

A deeper reason why spinors only do half of what one thought everything in the world should do, and at the same time a deeper answer to the question "Why spinors at all?", is the connection between $SU(2)$ and $SO(3)$, which we shall discuss now. As already remarked, both the three-dimensional rotation matrices $SO(3)$ and the complex rotation matrices $SU(2)$ [see (1.42)] are groups. To get to the bottom of the connection, we note that the rotation group $SO(3)$ is also a continuum, called a manifold in mathematics, and that makes it into what is called a Lie group.

The continuum character is easily seen as follows. Each rotation is entirely specified by a rotation axis in space and a rotation angle about that axis. One possible way to capture this mathematically, i.e., to provide a representation of this, can be given by considering the ball $B_0(\pi) \subset \mathbb{R}^3$ centred at 0 and with radius π. We choose a radial direction or ray, starting from 0 and intersecting the upper half of the ball, and mark on that ray the length $\alpha \leq \pi$. The ray defines the axis of rotation and α the rotation angle. The mirror image of this ray is interpreted as rotation through angle $-\alpha$ around that same axis. In this way any rotation in $\mathbb{R}^3$ would be represented in a unique manner, if it weren't for the pairs constituted by any pole and its antipode, corresponding to π and $-\pi$. What should we do about them?

Clearly, rotating a vector through π or $-\pi$ around a chosen axis yields the same result. This means that we must identify the antipodes, i.e., we must view them as being the same point. Mathematically, that is easily done, but it is not easy to picture. The message which comes with identifications is in general that they create topological complications. Take, for example, a path from one pole of the sphere to its antipode. That path is a closed path under the above identification. Such a path cannot be deformed to a point without being broken somewhere. In mathematical terminology, it cannot be continuously deformed to a point (try it and see!). Put another way, the path is not *null homotopic* (it could be useful here to review the notion of homotopy from the vector analysis class). If we now adjoin a second path from antipode to pole, which is then also closed, we obtain a doubly closed path. We can now deform the adjoined path continuously so that it coincides geometrically with the first one, but running backwards through it, and hence undoing all rotations which define the first path and resulting in a zero net rotation. So the doubly closed path is a null homotopic path.

A manifold containing paths that are not null homotopic is said to be *multiply connected* or *not simply connected*. The plane $\mathbb{R}^2$ is simply connected, while the torus (the surface of a donut) is not. We shall return to this idea in Sect. 4.4 and Fig. 4.3 in the context of identical particles. Hence the manifold $SO(3)$ is not simply connected.

Now, a simply connected manifold which can be projected onto a not simply connected manifold is called a (universal) covering. The message is here is that life is easy on the covering and not so easy on the not simply connected manifold below it. So we aim to do our physics as much as possible on the covering. The universal covering of $SO(3)$ is in fact the Lie group $SU(2)$ with elements

$$U = \begin{pmatrix} a & b \\ -b^* & a* \end{pmatrix}, \quad a = \alpha_1 + i\alpha_2, \quad b = \beta_1 + i\beta_2, \quad \alpha_1^2 + \alpha_2^2 + \beta_1^2 + \beta_2^2 = 1.$$

The rightmost equality is the determinant condition, which means that the four real numbers specifying the matrix also determine the three-dimensional unit sphere S^3 centred at 0. Being a spherical surface, it is a manifold, and what's more, it is simply connected.

Therefore, $SU(2)$ turns out to be topologically equivalent (or diffeomorphic) to a three-dimensional sphere S^3 and is thus simply connected. $SU(2)$ is a double covering of $SO(3)$: the $SU(2)$ matrices U and $-U$ yield the same rotation in $\mathbb{R}^3$ (see Remark 1.6) and the identification of a pole and its antipode on the $SO(3)$ manifold is thus resolved into two corresponding $SU(2)$ elements with opposite sign. For those who find this too mathematical, (1.40) and Remark 1.6 should be good enough.

The topological property of not being simply connected motivates the search for a universal covering which is mathematically simpler than the underlying multiply connected manifold and—perhaps with some surprise—we eventually find that the universal covering plays a fundamental role in physics. That by itself should of course convince nobody that spinors must appear in the description of quantum phenomena. The fact that they do was established by Dirac. More will be said about this in Chap. 11.

In any case, we can understand why the name "spin" is not completely off target. Spinors allow a representation of $SU(2)$, which is the universal covering group of $SO(3)$, the three-dimensional rotation group. We can also refer to the famous Noether theorem. Corresponding to the rotational invariance of the wave equation for a closed system (e.g., the Pauli equation), this implies the conservation of the total spin of that system. But it would nevertheless be incorrect to think of a spinor as an object which rotates.

1.9 Hilbert Space and Observables

Wave functions to which Born's statistical interpretation applies must be square integrable:

$$\int_{\mathbb{R}^{3N}} |\psi|^2(q)\, \mathrm{d}^{3N}q < \infty.$$

The integral here is the Lebesgue integral (Henri Lebesgue 1875–1941). The advantage of Lebesgue integration is that, under mild conditions, limits of Lebesgue integrable functions are again Lebesgue integrable. The Lebesgue integral can be used to define a norm and the vector space of Lebesgue integrable functions turns out to be a Banach space, i.e., a normed space which is complete in the sense that every Cauchy sequence of elements in the space has a limit in the space.

One consequence is that the vector space of wave functions which are square integrable is a Hilbert space[23]:

$$L^2(\mathbb{R}^n, d^n q) := \left\{ \psi : \mathbb{R}^n \to \mathbb{C} \,\middle|\, \langle \psi | \psi \rangle := \int |\psi(q)|^2 \, d^n q < \infty \right\}.$$

A Hilbert space is a Banach space, with the norm induced by a scalar product. In our case, the scalar product is given by

$$\langle \varphi | \psi \rangle := \int_{\mathbb{R}^n} \varphi^*(q) \psi(q) \, d^n q$$

and the induced L^2-norm is $\|\psi\| = \sqrt{\int |\psi(q)|^2 d^n q}$. All this is based on the construction of the Lebesgue measure and the Lebesgue integral and is part of any decent education in analysis.

L^2 has a countably infinite basis, where the notion of basis is different from that usually considered in courses on linear algebra. The difference is that, in an infinite-dimensional Hilbert space, although we can once again represent any vector in an orthonormal basis, this will in general require infinitely many coordinates. Letting ϕ_k, $k \in \mathbb{N}$, be an orthonormal basis, so that $\langle \phi_k | \phi_l \rangle = \delta_{k,l}$, then any vector is represented by an infinite sequence that is convergent in the L^2-sense:

$$\varphi = \sum_k \langle \phi_k | \varphi \rangle \phi_k . \tag{1.44}$$

The coordinates

$$c_k = \langle \phi_k | \varphi \rangle$$

are square summable, i.e.,

$$\sum_k |c_k|^2 < \infty .$$

The convergence of (1.44) is shown using the Cauchy property and completeness.

Given an orthonormal basis, the set of the coordinate vectors with countably infinitely many components is also a Hilbert space, denoted by l^2. It was the space of square summable vectors which David Hilbert (1862–1943) first introduced and which was then extended to the general Hilbert space. Those who don't like to think in abstract mathematical terms or who are not particularly trained to do so can think of vectors in a coordinate representation. For example when asked "What is a vector?", we may give the answer $(x_1, \ldots, x_n)$. And likewise in answer to the

---

[23] We shall often simply write L^2 for $L^2(\mathbb{R}^n, d^n q)$.

question "What is a wave function?", we might reply $(c_1, \ldots, c_n, \ldots)$. But, and this is one of the take-home lessons from a course in linear algebra, we must always specify the orthonormal basis for which these are the coordinates.

When Werner Heisenberg (1901–1976) invented his matrix mechanics as a description of quantum phenomena, he actually had his matrices act on l^2. Of course, the matrices had then to have infinitely many rows and columns, and these are, by the way, mathematically well defined objects, namely linear operators. The Hilbert space formalism explains why Heisenberg's matrix mechanics (l^2) and Schrödinger's wave mechanics (L^2) are equivalent descriptions.

As in matrix mechanics, we consider linear operators on the Hilbert space L^2, and in particular self-adjoint operators, which can be considered as generalisations of symmetric matrices, some of which represent so-called observables in quantum mechanics. An observable is an object which is associated with a measurement experiment. An example is the momentum observable (or momentum operator) which we already derived below Eq. (1.13). We also introduce the position operator $\hat{\mathbf{X}}$ of a particle, although the definition may look somewhat simple-minded:

$$\hat{\mathbf{X}}\psi(\mathbf{x}) := \mathbf{x}\psi(\mathbf{x}).\tag{1.45}$$

For the position operator, quite analogously to what happens for the momentum operator [see (1.13) and below], the expected value in a "measurement of the position operator" for a given wave function can be computed by

$$\langle\psi|\hat{\mathbf{X}}\psi\rangle = \int_{\mathbb{R}^3} \mathbf{x}\,\psi^*(x)\psi(x)\,\mathrm{d}^3x .$$

Similarly,

$$\langle\psi|\hat{\mathbf{X}}^2\psi\rangle = \int_{\mathbb{R}^3} \mathbf{x}^2\psi^*(x)\psi(x)\,\mathrm{d}^3x .$$

Now that we have introduced these objects we may as well rephrase (1.24) in terms of operators. To do this we introduce a time-dependent operator, the so-called Heisenberg operator. Recall the Schrödinger equation (1.21). As physicists we have no problem writing the solution as

$$\psi(\mathbf{x}, t) = \exp\left(-\mathrm{i}\frac{t}{\hbar}H\right)\psi(\mathbf{x}) =: U(t)\psi(\mathbf{x}),$$

and because H is Hermitian (which means the same as self-adjoint), it is clear that $U(t)^+U(t) = \mathbf{1}$, i.e., it is unitary. We now define

$$\langle\psi|\hat{\mathbf{X}}(t)\psi\rangle := \langle\psi(\mathbf{x}, t)|\hat{\mathbf{X}}\psi(\mathbf{x}, t)\rangle = \langle\psi|U^+(t)\hat{\mathbf{X}}U(t)\psi\rangle ,$$

or simply $\hat{\mathbf{X}}(t) := U^{+}(t)\hat{\mathbf{X}}U(t)$. Then, omitting the scalar product, (1.24) yields

$$m\frac{\mathrm{d}^2\hat{\mathbf{X}}}{\mathrm{d}t^2} = -\nabla V(\hat{\mathbf{X}}) . \tag{1.46}$$

This implies that the Heisenberg position operator satisfies the classical equation of motion, something that is actually true for other Heisenberg operators, e.g., momentum. This is quite nice, but deceptively so, since operators are not physical entities moving in space. So we have to keep in mind all the cautionary remarks that were made in the section about the classical limit (Sect. 1.6).

We shall elaborate on the meaning of operators in Chap. 7. It is often the case that physics students with high concern for mathematics think that the "problems of quantum mechanics" can simply be solved by good clean mathematics! For example, the position and momentum operator are unbounded linear operators. That means that their domain of definition is not the whole of L^2. We need only glance at (1.45) to see that there exist wave functions for which the right-hand side of (1.45) is not square integrable (find some!). So some students may develop the idea that, if all this were cleanly stated and observed, then quantum physics would be just fine. But it is not. The famous debate about quantum physics has nothing to do with messy mathematics.

The Measurement Problem

2

It is difficult to believe that this description is complete. It seems to make the world quite nebulous unless somebody, like a mouse, is looking at it.
                          Albert Einstein, from his last lecture at Princeton on 14 April 1954

We have to talk about Schrödinger's cat. Not just because Erwin Schrödinger found such a memorable way to illustrate his paradox. More importantly, his criticism hits the nail right on the head. Schrödinger formulates the so-called *measurement problem* of quantum mechanics, and this measurement problem shows why the naive understanding of quantum theory is not just unsatisfactory but completely untenable. Every precise formulation of quantum mechanics must therefore be judged by whether and how it resolves the cat paradox. We quote the famous section from Schrödinger's article *The present situation in quantum mechanics*[1]:

One can even set up quite ridiculous cases. A cat is penned up in a steel chamber, along with the following device (which must be secured against direct interference by the cat): in a Geiger counter there is a tiny bit of radioactive substance, so small, that perhaps in the course of the hour one of the atoms decays, but also, with equal probability, perhaps none; if it happens, the counter tube discharges and through a relay releases a hammer which shatters a small flask of hydrocyanic acid. If one has left this entire system to itself for an hour, one would say that the cat still lives if meanwhile no atom has decayed. The psi-function of the entire system would express this by having in it the living and dead cat (pardon the expression) mixed or smeared out in equal parts.

It is typical of these cases that an indeterminacy originally restricted to the atomic domain becomes transformed into macroscopic indeterminacy, which can then be resolved by direct observation. That prevents us from so naively accepting as valid a "blurred model" [as an image of] reality. In itself it would not embody anything unclear or contradictory. There is a difference between a shaky or out-of-focus photograph and a snapshot of clouds and fog banks.

---

[1]Die Naturwissenschaften **23** (48), 807–812 (1935). Translation by John D. Trimmer, originally published in Proceedings of the American Philosophical Society **124**, 323–38 (1980).

© Springer Nature Switzerland AG 2020
D. Dürr, D. Lazarovici, *Understanding Quantum Mechanics*,
https://doi.org/10.1007/978-3-030-40068-2_2

A first reading of these lines may leave the reader somewhat confused. What Schrödinger says is exactly right and exactly what has to be said, but he has not been too considerate with the reader. He doesn't really help us get the point, especially if we are used to skimming over written text to extract information quickly. But Schrödinger was famous and he could afford to say things as they are and wait to be understood. Unfortunately, though, things went wrong in this case. The cat story has become folklore, but absurd discussions have distorted the content and also the genesis of the problem beyond recognition. Schrödinger presented his argument as a *reductio ad absurdum* of the assertion that standard quantum mechanics provides a complete description of nature: "That prevents us from so naively accepting as valid a 'blurred model' as an image of reality." Nowadays, the poor cat, equal parts dead and alive, is often cited as an example of how bizarre nature really is according to quantum mechanics, and some physicists even seem to take pride in being able to deal with such a crazy description of reality. Somewhere along the way, the seriousness with which physics had once been practised has clearly been lost.

So let's slow down and start over. The problem with quantum mechanics is the following: there is only one equation and one quantity specifying the contents of the theory, viz., the Schrödinger equation and the associated wave function, and they do not describe phenomena as we perceive them. This can be seen in different ways, for instance as follows. Suppose a system is described by a linear combination of wave functions φ_1 and φ_2 and a piece of apparatus can display either "φ_1" or "φ_2" by interacting with the system. This, in brief, is what we call a "measurement". In principle, this apparatus must also have a quantum mechanical description. After all, we conceive of the measuring apparatus as consisting of atoms and molecules, and if all these atoms and molecules are described by a wave function, then this wave function must also provide a quantum mechanical description of the apparatus as a whole. This means that the apparatus has states Ψ_1 and Ψ_2 corresponding to pointer positions which we call "1" and "2", i.e., these are wave functions that have disjoint support (no overlap) in configuration space (see Figs. 1.1 and 1.2), and it also has a prepared state Ψ_0 such that

$$\varphi_i \Psi_0 \quad \xrightarrow{\text{Schrödinger evolution}} \quad \varphi_i \Psi_i \, . \tag{2.1}$$

The Schrödinger time evolution (2.1), however, is linear. Therefore, a system wave function

$$\varphi = c_1\varphi_1 + c_2\varphi_2 \, , \qquad c_1, c_2 \in \mathbb{C}, \qquad |c_1|^2 + |c_2|^2 = 1 \, ,$$

leads to

$$\varphi\Psi_0 = (c_1\varphi_1 + c_2\varphi_2)\Psi_0 \quad \xrightarrow{\text{Schrödinger evolution}} \quad c_1\varphi_1\Psi_1 + c_2\varphi_2\Psi_2 \, . \tag{2.2}$$

This is an absurd result. The superposition

$$c_1 \varphi_1 \Psi_1 + c_2 \varphi_2 \Psi_2 \tag{2.3}$$

describes an entangled state between system and apparatus, in which the pointer position seems to indicate both "1" and "2" at the same time. In Schrödinger's cat experiment, φ_1 would describe the already decayed atom and φ_2 the not yet decayed atom. But this would mean that (2.3) describes a state with a dead cat and a live cat.

Thus, if we insist that the result of the measurement is either "1" or "2" but not both at the same time, we have the following situation. Either the wave function of the system after measurement is not (2.3), in which case the Schrödinger equation is not correct, or at least not always; or the wave function of the system is indeed (2.3), but this wave function does not provide a complete description of the physical situation. In this case, we are missing precisely those variables that make the difference between a pointer pointing to the left and a pointer pointing to the right— the difference between a dead cat and a live cat.

It is sometimes explained that the wave function should not be taken seriously in this sense, and that only the statistical interpretation according to Born's rule is significant. Then, according to Born's rule, the result of the measurement is "1" with probability $|c_1|^2$ or "2" with probability $|c_2|^2$. It is this statistical law that is confirmed experimentally to great accuracy. Fair enough. But pointing to Born's rule does not avoid the measurement problem. The Schrödinger equation is a deterministic equation, and according to this equation, the wave function at the end of the experiment is always (2.3). If this wave function provides a complete description of system and apparatus, the outcome of the measurement is always the same. In other words, if the wave function (2.3) provides a complete description of system and apparatus, it cannot on some occasions describe a measurement device whose pointer points to the left and on other occasions a measurement device whose pointer points to the right.

So how is the statistical interpretation to be understood? There are two possibilities. If we mean that $|c_i|^2$ is the probability that the wave function of the system + apparatus after the measurement is $\varphi_i \Psi_i$, then the Schrödinger equation cannot always be valid since, according to the Schrödinger equation, the wave function will be (2.3). But if we mean that the wave function of the system is (2.3) and this wave function only gives us the probability distribution for the actual state of the system, then it is clear that the actual state of the system is not described by the wave function alone, and we are missing precisely those physical variables whose probability distribution the wave function is supposed to provide. The dilemma is thus the same as before.

2.1   The Orthodox Answer

In order to maintain the completeness of the quantum mechanical description no matter what, Werner Heisenberg, John von Neumann (1903–1957), and others introduced an additional postulate into the theory. In the process of "measurement" or "observation", they postulated, the Schrödinger time development is suspended and replaced by a random dynamic which reduces the superposition (2.3) with probability $|c_i|^2$ to the wave function $\varphi_i \Psi_i$. However, in contrast to the Schrödinger evolution, this new dynamic, the *collapse of the wave function*, was not supposed to be described by a precise mathematical law. It was introduced *ad hoc*, as a property of "the observer".

Indeed, it is precisely for this reason that the observer assumes a central role in the theory, as the subject whose act of measurement or observation brings about the physical facts. Wolfgang Pauli described this as an "act of creation lying outside the laws of nature". However, it is not even these esoteric traits of the Copenhagen school that prevent us from accepting the collapse postulate as part of a precise physical theory, but simply what John Bell described as its "unprofessional vagueness". We now have two contradictory dynamics for the wave function—when exactly does one or the other apply? When exactly is a physical process considered to be a "measurement"? And what distinguishes an "observer" from a molecule, or a cat, or the pointer of a piece of apparatus? Here is Bell once again, in his article *Against the measurement*:

> It would seem that the theory is exclusively concerned about 'results of measurement', and has nothing to say about anything else. What exactly qualifies some physical systems to play the role of 'measurer'? Was the wavefunction of the world waiting to jump [collapse] for thousands of millions of years until a single-celled living creature appeared? Or did it have to wait a little longer, for some better qualified system ... with a Ph.D.? If the theory is to apply to anything but highly idealised laboratory operations, are we not obliged to admit that more or less 'measurement-like' processes are going on more or less all the time, more or less everywhere? Do we not have jumping then all the time? (J.S. Bell, "Against the measurement". In: *Speakable and Unspeakable in Quantum Mechanics*, Cambridge University Press, 2nd edn. 2004, p. 216.)

Ironically, discussions about whether the cat has enough consciousness to trigger the collapse of the wave function (or whether it requires a human observer to seal its fate) have been going on for decades. Some people are still discussing this today. The reason why hardly anyone is still defending the Copenhagen interpretation is that at some point, physicists wanted to take quantum theory seriously as an objective description of nature. In particular, when cosmology was established as an important discipline in physics, quantum theory was supposed to tell us something about the creation of matter in the early universe, possibly even about the evolution of the universe itself. And shortly after the Big Bang, more than 13 billion years ago, there was certainly nothing and no-one around that would qualify as an observer.

2.2 Solutions to the Measurement Problem

The old Copenhagen quantum mechanics did not provide a serious solution to the measurement problem. Now let's see what the serious solutions are. Following Tim Maudlin,[2] a precise and general formulation of the measurement problem can be given as follows. There are three principles or propositions that a naive reading of the theory seems to assume, and these three principles together are logically inconsistent, i.e., they cannot all be true:

1. The wave function of a system provides a complete description of its physical state.
2. The time evolution of this wave function always follows a linear (Schrödinger) equation.
3. Measurements (usually) have unique outcomes.

The contradiction following from these three assumptions was derived above: it is precisely Schrödinger's cat paradox. If the wave function obeys a linear Schrödinger equation, the measurement procedure (2.2) results in a macroscopic superposition (2.3). If (2.3) provides a complete description of the physical state of the measurement device, the outcome of the measurement cannot be unique. Thus, $(1) \wedge (2) \Rightarrow \neg(3)$. In other words, (1), (2), and (3) are logically incompatible. Any consistent formulation of quantum mechanics must negate at least one of them.

2.2.1 The Negation of (1) Leads to Bohmian Mechanics

If we deny that the wave function provides a complete description of the physical state of a system, we have to name the missing pieces that turn an incomplete description into a complete one. As we will later see, the quantum phenomena thereby commit us to a radical minimalism. It is neither possible nor desirable to assume that all quantities ("observables") that quantum mechanics usually speaks about have definite values that are part of the state description. We have to pose the question differently: If quantum mechanics is not about the wave function, what is it about? Or if it is not *only* about the wave function, what else is it about?

David Bohm's response was straightforward: Bohmian mechanics is a theory about the motions of point particles. The "quantum observables" and their statistics follow from a statistical analysis of these particle motions. The complete description of the physical state of a system is thus given by its wave function *and* the positions of the particles constituting the system. For an N-particle system, this is a pair (ψ, Q), where $Q \in \mathbb{R}^{3N}$ describes the actual particle configuration in three-dimensional space. The role of the wave function ψ is first and foremost to guide the motion of the particles. This is expressed by a precise mathematical law which

---

[2]T. Maudlin, Three Measurement Problems. Topoi **14**, 7–15 (1995).

involves the wave function. The measurement problem is solved because every system has a well-defined configuration at all times, given by the positions of its particles. The wave function of a measurement device (for example) may be in a superposition (2.3), but the actual configuration Q describes a pointer pointing *either* to the left *or* to the right.

2.2.2   The Negation of (2) Leads to Collapse Theories Such As GRW

If we insist on (1) and (3), then the measurement problem is a result of the linearity of the Schrödinger equation. The superposition principle is, however, crucial to the explanation of quantum phenomena—just think about the double slit experiment. So, how can we save the superposition principle for microscopic systems—for which these phenomena are observed—but avoid macroscopic superpositions such as "dead cat" + "live cat" which lead to the measurement problem? The relevant concepts are already there in standard quantum mechanics. We have the linear Schrödinger equation and the collapse of the wave function. The problem is that this collapse is not a precise physical law, but an *ad hoc* postulate relying on vague and ambiguous notions such as "measurements" and "observations".

In a precise quantum theory, the collapse of the wave function should also be described by a precise mathematical law. Ghirardi, Rimini and Weber were the first to propose such a law. In their *GRW theory*, the Schrödinger equation is replaced by a non-linear, stochastic equation that already contains the possibility of collapse. This law is such that each "particle" has a certain collapse probability which is so low that superpositions for small systems, with only a few particles, persist for a very long time. However, in a macroscopic system, consisting of billions of billions of particles, a collapse will almost certainly be triggered in a tiny fraction of a second. In this sense, a "measurement"—that is, the coupling of a macroscopic piece of apparatus to a microscopic system—can indeed cause the collapse of the wave function, but this collapse is now part of a precise fundamental law. The measurement problem is solved because the nonlinear time evolution destroys macroscopic superpositions such as that of "dead cat" and "live cat" on empirically relevant time scales.

2.2.3   The Negation of (3) Leads to the Many Worlds Theory

If we insist on (1) and (2), we have no other choice but to accept macroscopic superpositions such as those of "dead cat" and "live cat" as a consequence of quantum mechanics. One radical conclusion, generally attributed to Hugh Everett III, is that both parts of this wave function describe a real physical state. But then, wouldn't we observe two cats rather than one? Or maybe a cat in an absurd hybrid state of "dead" and "alive"? The answer is that we would not, because the superposition of the wave function doesn't stop with the cat. The superposition would come to include the experimenter herself, the laboratory, indeed the whole of the rest of the universe, and consequently everything would get caught up in the splitting described by (2.3).

When all is said and done, this description of nature thus comprises two "worlds", corresponding to the branches of the wave function: in one world, the radioactive atom has decayed, the cat is dead, and the experimenter is sad; in the other, the atom hasn't decayed, the cat is alive and well, and the experimenter takes the animal back home. Because of the linearity of the Schrödinger equation, the two "copies" of the experimenter can never interact, and decoherence makes it practically impossible to bring the dead cat and the living cat into interference. Thus, both worlds can exist in parallel, without observers in one world ever (directly) perceiving the other.

The measurement problem is solved by accepting its consequences and trying to reconcile them with our experience: measurements do not have unique outcomes. Instead, *all* possible outcomes are realized in different worlds corresponding to different branches of the wave function. However, since everything and everybody is involved in this branching, this does not in itself contradict empirical evidence.

2.3   Other Alternatives?

In the following, large parts of this book will be devoted to working out these three quantum theories in more detail and showing that they can indeed provide a deeper and clearer understanding of quantum mechanics. A final assessment will be given in the epilogue. Physical theories are never without alternatives. Yet, we have seen that an analysis of the measurement problem leads quite naturally to Bohmian mechanics, GRW, and Many Worlds as possible solutions.

Without doubt, it would be conceivable to add other ontological variables to the wave function than particle positions, but Bohmian mechanics is by far the simplest and most successful quantum theory following this approach. Similarly, there are other ways to modify the Schrödinger equation than the original proposal by Ghirardi, Rimini, and Weber, though all options that have been seriously considered (and, in fact, all options that are mathematically possible under certain assumptions) are generalizations or variations of the GRW theory. Finally, although a Many Worlds picture is essentially unavoidable if we insist on the completeness of the wave function as a state description and the linearity of its time evolution, there do exist different proposals on how to spell out the details of a Many Worlds theory.

There are, of course, countless other "interpretations" of quantum mechanics floating around in the literature. Some of them develop a distinct formalism, but many are mere attempts to discuss the measurement problem away. Addressing all of these proposals in detail would go beyond the scope of this book. But more importantly, it could suggest a false equivalence between theories at very different levels of soundness, maturity, and productivity.

Instead, the above formulation of the measurement problem should provide a useful scheme that the reader can employ to make her own judgements. For instance, some people may claim that their version of quantum mechanics violates none of the assumptions (1), (2), and (3). Then, we may readily infer that their version of quantum mechanics is not a sound physical theory. Others may effectively deny assumption (1), e.g., by saying "the wave function does not describe the actual state of the system but only my (incomplete) information about it." We should press these people to detail the quantities and laws describing the actual state of the system, i.e., the physical facts that they claim to have information about. In general, one will not receive a clear answer.

Of course, there will always be voices (even from otherwise reasonable people) insisting that it's simply not the role and purpose of physics to provide a coherent and objective description of nature. Since this is not a scientific position but a philosophical one, it can never be conclusively disproven. What can be conclusively disproven is the claim that quantum phenomena preclude any coherent, objective description of nature. The three "quantum theories without observers" presented in this book are a proof to the contrary.

2.4   The Measurement Problem and Born's Statistical Hypothesis

We carry out a simple computation demonstrating the application of Born's statistical hypothesis in Bohmian mechanics and GRW. The justification of this statistical hypothesis—a truly important and subtle issue—will be addressed later. Our focus here is on something else. We have seen above that the measurement problem of quantum mechanics is also manifested in an ambiguity about the meaning of Born's rule: the $|\psi|^2$ distribution gives a probability, but the probability *of what*? We need to convince ourselves that a precise formulation of quantum mechanics provides a precise answer to this question—although the answer may be different depending on the theory.

Our computation follows our discussion of the measurement process. Let the configuration space of the complete system be described by coordinates $\mathbf{q} = (\mathbf{x}, \mathbf{y})$, where $\mathbf{x} \in \mathbb{R}^m$ are the coordinates of the measured system and $\mathbf{y} \in \mathbb{R}^n$ those of the measurement apparatus. According to Born's rule, we have

$$\mathbb{P}(\text{pointing to } 1) = \int_{\text{supp } \Psi_1} |c_1 \varphi_1 \Psi_1 + c_2 \varphi_2 \Psi_2|^2 \, \mathrm{d}^m x \, \mathrm{d}^n y \tag{2.4}$$

$$= |c_1|^2 \int_{\text{supp } \Psi_1} |\varphi_1 \Psi_1|^2 \mathrm{d}^m x \, \mathrm{d}^n y$$

$$+ |c_2|^2 \int_{\text{supp } \Psi_1} |\varphi_2 \Psi_2|^2 \mathrm{d}^m x \, \mathrm{d}^n y$$

$$+ 2\mathrm{Re}\left[c_1 c_2 \int_{\text{supp } \Psi_1} (\varphi_1 \Psi_1)^* \varphi_2 \Psi_2 \mathrm{d}^m x \, \mathrm{d}^n y \right] \tag{2.5}$$

$$\approx |c_1|^2 \int |\varphi_1 \Psi_1|^2 \mathrm{d}^m x \, \mathrm{d}^n y = |c_1|^2 \, . \tag{2.6}$$

Note here that, since the supports of the two pointer wave functions are disjoint (or nearly so), the term (2.5) is zero. The probability of the outcome "1" is thus $|c_1|^2$ and the probability of the outcome "2" is $|c_2|^2$, just as the rules of textbook quantum mechanics would suggest. What exactly have we calculated, though?

The answer given by Bohmian mechanics corresponds to our usual way of speaking: Born's rule provides the probability distribution for the particle positions. $|c_1|^2$ is thus the probability that, at the end of the measurement process, the pointer—consisting of a great number of particles—actually points left, to the result "1".

In the GRW theory, the same computation is interpreted differently. Here, Born's rule provides first and foremost a probability distribution for the center of the collapse. More simply put, $|c_1|^2$ is first and foremost the probability that (2.3) collapses onto a wave function that is localized in the support of Ψ_1, i.e., on configurations corresponding to a pointer pointing left, to the result "1".

The interpretation of Born's rule in the Many Worlds theory is difficult. Here, it doesn't really make sense to say that $|c_1|^2$ is the probability that the measurement outcome "1" occurs, because *all* possible outcomes occur with certainty, but in different worlds. In Chap. 6, we will address this problem in greater detail. For now, a common, though somewhat evasive response is that the Born probabilities have to be interpreted as subjective probabilities. That is, after the measurement (and the splitting of the world) has occurred, but before we know the result in our world, we should have a credence of $|c_1|^2$ that we will find ourselves in a world in which the pointer points to the left.

Finally, we may ask about the meaning of Born's rule in the orthodox (Copenhagen) quantum theory. We might perhaps say that the above computation describes a "position measurement" of the pointer. Then, $|c_1|^2$ is the probability that the pointer points left, to "1", if we look at it, but decidedly *not* the probability that the pointer points left in the case when nobody looks. Alternatively, we could forbid the computation altogether and insist that the probabilities must come from an observable operator (for instance a "cat-aliveness operator"). Finally, the most orthodox answer of all is that the computation is forbidden, because a measurement device is too big to have a wave function. If all this doesn't sound too serious, that's because it isn't. On the other hand, it is the kind of talk that has surrounded quantum mechanics for decades.

In any case, a basic idea still found in textbook quantum mechanics is that observables (self-adjoint operators) relate the quantum state (wave function) directly to the measurement statistics of observable quantities. We have already hinted at why this is a bad idea. A measurement is itself a complex physical process, and only an analysis of the theory can tell us what the observable quantities are and how they can be measured in experiment. Taking observations as fundamental and trying to build a theory on that basis will never work. Indeed, it leads to disaster, more precisely, to the various "quantum paradoxes" that are still the subject of endless unnecessary debates. In Chap. 7, we will discuss what the role of the observable operators really is. They arise quite naturally from a precise formulation of quantum mechanics, but they are not, of course, fundamental.

2.5   Decoherence

In the present section it will be helpful to keep Figs. 1.1 and 1.2 in mind. Our discussion of the measurement process has naturally involved an important concept that was already mentioned in Chap. 1: *decoherence*. Decoherence means that different parts of the wave function lose their ability to interfere, often due to entanglement with a macroscopic environment. With regard to Eq. (2.2), this should be understood in the sense that Ψ_1 and Ψ_2 are concentrated on well-separated regions of configuration space and that the many degrees of freedom of a macroscopic wave function make it practically impossible to bring them into overlap. Put another way, in three-dimensional space, there are very few ways to pass by each other, viz., left/right, above/below, in front/behind, while the configuration space of a macroscopic system has roughly 10^{24} dimensions, meaning countless possibilities for the peaks of the wave packets to miss each other.

Decoherence is actually happening everywhere and all the time, unless one takes special precautions to isolate a system from the environment. A common measurement, however, must basically decohere the measured system by definition if the measurement is supposed to be informative. A measurement device is a macroscopic system which interacts with a microscopic system in such a way that it will end up in one of several, macroscopically discernible configurations that indicate a particular measurement result. And "macroscopically discernible" means precisely that the different "pointer states" must be concentrated in well-separated regions of configuration space.

It is sometimes claimed that decoherence alone provides the solution to the measurement problem. This is simply wrong. The measurement problem is thereby misunderstood as the problem of showing that the wave functions in (2.3) do indeed have more or less disjoint supports, i.e., that the step from (2.4) to (2.6) is justified.

But the fact that decoherence doesn't solve the measurement problem should already be evident from the fact that coherence—the ability of the superposed wave packets to interfere—plays no role in its formulation. The point of Schrödinger's cat argument is that the wave functions of a dead cat and a living cat are there at the same time, and that they are equally "real", *not* that they overlap in configuration space and produce interference. This is something that Schrödinger understood very well. As long as we insist that the wave function provides a complete description of the physical state, the result of the measurement is still "dead cat AND live cat", not "dead cat OR live cat".

When interference phenomena (or the lack thereof) are discussed in the literature, we usually consider not the wave function itself but the so-called *statistical operator* or *density matrix*, since this formalism allows us to represent both the *mixture*

$$\varrho_M = |c_1|^2 \, |\varphi_1\rangle |\Psi_1\rangle \langle\Psi_1| \langle\varphi_1| + |c_2|^2 \, |\varphi_2\rangle |\Psi_2\rangle \langle\Psi_2| \langle\varphi_2| \qquad (2.7)$$

and the *pure state*

$$\varrho = |c_1\varphi_1\Psi_1 + c_2\varphi_2\Psi_2\rangle\langle c_1\psi_1\varphi_1 + c_2\Psi_2\varphi_2|. \qquad (2.8)$$

This provides a new way of speaking but no new insights. Decoherence in the sense of separation on configuration space is now manifested in the way that the *off-diagonal elements* $c_1 c_2|\varphi_1\Psi_1\rangle\langle\Psi_2\varphi_2|$ and $c_1 c_2|\varphi_2\Psi_2\rangle\langle\Psi_1\varphi_1|$ are almost zero in the pure state (2.8), which corresponds to the superpositions (2.3). The convergence $\varrho \to \varrho_M$ can then be proven in the thermodynamic limit by averaging over (referred to as tracing out) the degrees of freedom of the environment[3].

The punchline is now that a density matrix of the form (2.7) is also what an experimenter would use to describe a series of measurements for which the state of the system cannot be reliably prepared. The prepared system has the wave function φ_1 with probability p_1 or the wave function φ_2 with probability p_2, but she doesn't know exactly which in each individual case. The statistics of the experiment can then be described by the "statistical mixture" $\varrho_M = p_1|\varphi_1\rangle\langle\varphi_1| + p_2|\varphi_2\rangle\langle\varphi_2|$.

This pseudo-solution of the measurement problem now claims that the limit $\varrho \to \varrho_M$ justifies the same ignorance interpretation for the decohered state (2.8). But this is just a sleight of logic, because a thermodynamic limit cannot turn an AND into an OR. After all, the linearity of the Schrödinger evolution is never suspended, so the wave functions of "dead cat" and "live cat" are still both there in the final state, no matter how big the environment is and how well the two wave packets are separated. This is exactly what Schrödinger meant when he wrote: "There is a difference between a shaky or out-of-focus photograph and a snapshot of clouds and fog banks."

2.6 The Ontology of Quantum Mechanics

We have formulated the measurement problem as a trilemma—three logically incompatible assumptions, at least one of which must go. This has led us to Bohmian mechanics, GRW, and the Many Worlds theory as consistent solutions. At the end of the day, however, we want more than a consistent mathematical formalism. We want a coherent description of nature. What is matter—a cat, or a table, or a piece of measurement apparatus—actually made of? What is it that actually goes through the slits in the double-slit experiment and hits the screen to create an interference pattern? Those are questions about the *ontology* of quantum mechanics that we have briefly addressed in Chap. 1. In fact, the lack of a clear ontology in orthodox quantum mechanics is the real root of the measurement problem (and many other problems). If the ontology is clear—if it is clear what the fundamental entities in nature are that the theory seeks to describe—there can't be any paradoxes.

---

[3] See, e.g., E. Joos, H.D. Zeh, C. Kiefer, D. Giulini, J. Kupsch, and I.-O. Stamatescu, *Decoherence and the Appearance of a Classical World in Quantum Theory*. Springer, 2003.

This realization leads to a dilemma that cuts across the trilemma of the measurement problem: can the ontology of quantum mechanics consist of the wave function and the wave function alone, or do we have to add additional ontological variables in order to describe nature? The Many Worlds theory takes the first route and tries to develop a description of physical reality in terms of the wave function alone. The task is then to explain the connection between the wave function—a complex "field" on the high-dimensional configuration space—and stuff in three-dimensional space. Certainly, wave functions cannot make up cats, and tables, and measurement devices in the same way particles could. We will address this challenge in detail in Chap. 6.

In contrast, Bohmian mechanics postulates point particles as the ontology of quantum mechanics. By describing how particles move—and how they come together to form tables, and cats, and measurement devices—the theory describes what's going on in the physical world. The wave function, for its part, is understood primarily through its role in the dynamics of the particles—and features also in their statistical description, for reasons we will explain in Chap. 4.

Finally, in the GRW theory, both routes are actively pursued in the literature. GRW can be understood as a theory about the wave function. The spontaneous collapse then avoids having to postulate many worlds, but we encounter the same difficulty regarding the connection with our everyday experience as in the Many Worlds theory. There is still a difference between a table in three-dimensional space and a wave function concentrated on table configurations in the high-dimensional configuration space. However, it is also possible to conceive of GRW with an ontology of localized entities in three-dimensional physical space. These entities are not particles moving on continuous trajectories but either "matter flashes" associated with the collapse events or continuous "mass densities" defined in terms of the wave function. We will discuss this in more detail in Chap. 5.[4]

What is important to appreciate here is that the question *What is the theory about?* is absolutely fundamental in order to understand a theory and what it tells us about nature. And it is a *physical* question that the theory itself must answer, not a philosophical question that can be left open to "interpretation".

---

[4]Theoretically, it is also possible to do Many Worlds with a local ontology or Bohmian mechanics on configuration space, but in contrast to the different versions of GRW, those are rather exotic and largely academic proposals that tend to undermine the appeal of the original theories.

Chance in Physics

3

Assuming the success of efforts to accomplish a complete physical description, the statistical quantum theory would, within the framework of future physics, take an approximately analogous position to the statistical mechanics within the framework of classical mechanics. I am rather firmly convinced that the development of theoretical physics will be of this type; but the path will be lengthy and difficult.

Albert Einstein[1]

All quantum theories have one thing in common: they agree that chance (or randomness) should feel at home in the theory. It is sometimes claimed that quantum randomness, or equivalently quantum probability, is irreducible, by which it is meant that there is just no way out of it. Quantum mechanical probability is expressed in Born's statistical interpretation of the wave function and this is frequently taken as an axiom. Then nothing more can be said, unless one feels unhappy about the axiom. And indeed, there are reasons for not being happy about it.

For one thing, axioms usually carry an air of obvious truth about them. But probability is in itself a problematic notion (what exactly is probability?), suggesting that an axiom based on this notion will likewise be problematic. Moreover, apart from the so-called collapse theory, which does indeed involve an irreducible notion of chance, the Many Worlds theory or Bohmian mechanics are obviously deterministic, since the Schrödinger equation contains no random elements, and neither does the law of motion for the particles in Bohmian mechanics. Where then does randomness come from in such theories? What does a statistical hypothesis like Born's actually mean? To which systems does it apply? And what is the often used phrase "distributed like ..." supposed to mean?

---

[1]From: "Remarks concerning the essays brought together in this co-operative volume". In P.A. Schilpp (Ed.), *Albert Einstein Philosopher–Scientist*. MJF books, New York, 1949, p. 672.

© Springer Nature Switzerland AG 2020

D. Dürr, D. Lazarovici, *Understanding Quantum Mechanics*,

https://doi.org/10.1007/978-3-030-40068-2_3

It is noteworthy that these questions were (naturally) already raised in classical physics, and since they have been answered within that realm, we shall begin by re-examining classical physics to understand how Ludwig Boltzmann (1844–1906) dealt with them. We shall then see, and show in later chapters, that the very same reasoning can be carried over to quantum theory. In classical physics, the world only appears to be random, and the same is true in quantum theory, apart from the above-mentioned collapse theory. The basic insight is that the nebulous notion of probability needs to be replaced by the intuitive and easily accessible notion of *typicality*. Once this has been understood, everything becomes clear.

3.1   Typicality

When we toss a "fair" coin, why is the probability $1/2$ for both heads and tails? There are many possible correct answers. In particular, Laplace's [Pierre-Simon Laplace (1749–1827)] *principle of insufficient reason* will often be given: there are only two sides of the coin and both have equal rights to appear, whence the Laplace probability is set to $1/2$ for heads and $1/2$ for tails. In other words there is no sufficient reason to prefer one side over the other. But there is also a more profound idea which at first looks unrelated to the Laplace position.

We all agree that if we toss a coin many times the relative frequency of heads and tails will be close to the value $1/2$. But why is that so? Why is there this law-like behaviour in an otherwise random sequence of heads and tails? The new kind of law is called the *law of large numbers*. This law always felt rather strange, because chance behaviour seems to be the complete opposite of law-like behaviour. But the new law is not of the usual kind. It differs from let's say a law of physics. It is a new law which does not always hold, but only *typically*. This means that it is not necessarily true. The coin-tossing sequence could turn out differently, for example only $1/4$ heads and $3/4$ tails. But typicality wins out through overwhelming numbers. What we mean by that is shown in the following example.

Consider a coin-tossing sequence of length 1000, i.e., we toss a coin 1000 times, this representing the "large number" in the law of large numbers. The sequence can be viewed as a sequence of 0's and 1's of length 1000, where 0 stands for heads and 1 one for tails. The reader should check the answers to the following questions:

1. What is the total number of possible sequences of 0's and 1's of length 1000? It's simply 2^{1000}.
2. What is the number of possible sequences of length $n = 1000$ that contain exactly k heads? That is not so easy to find. But with some knowledge of high school mathematics and some reflection, we find that it is given by the binomial coefficient $\binom{n}{k}$ with $n = 1000$, that is $\binom{1000}{k}$.
3. What is the total number of possible sequences of length $n = 1000$ when $k \approx 500$ (equal distribution) and when k is very different from 500 (unequal distribution)? The answer can be read from Table 3.1.

Table 3.1 Absolute and relative numbers of sequences of 0's and 1's of length $n = 1000$ with exactly k heads

k	100	200	300	400	450	480	500
$\binom{1000}{k}$	10^{139}	10^{215}	10^{263}	10^{290}	10^{297}	10^{299}	10^{299}
$\binom{1000}{k}/2^{1000}$	$\frac{1}{10^{161}}$	$\frac{1}{10^{85}}$	$\frac{1}{10^{37}}$	$\frac{1}{10^{11}}$	$\frac{1}{10^4}$	$\frac{1}{100}$	$\frac{1}{40}$

Note that $\binom{n}{k}$ is symmetric about $n/2$. The given values are approximate

Remarkably, $\binom{1000}{300}$ differs from $\binom{1000}{500}$ by a factor of 10^{36}, which means that there are vastly more sequences close to the equal distribution. Compare this with the estimated age of the universe, which is 4×10^{17} s. This means that, even if you were able to produce a coin-tossing sequence of length 1000 every second, you would only produce $\sim 10^{17}$ sequences! The take-home lesson is this: the number of sequences of 0's and 1's of length 1000 which have roughly the same number of 0's and 1's is *overwhelmingly greater* than the number of sequences with notable differences in the numbers of 0's and 1's. What "notably different" means can also be seen from the table. Look at the bottom line of the table, showing the relative numbers: sequences with fewer than 450 1's (or 0's) contribute almost nothing to the total number of sequences. In general (and this is another take-home lesson), the sequences for which the difference in the number of 0's and 1's is at most of the order $\sqrt{n}$ (for sequences of length n) constitute *almost all possible sequences*, i.e., they are *typical*. This is called the $\sqrt{n}$ law.

The law of large numbers, which can be proven mathematically, says the obvious thing for the coin-tossing experiment, namely that there is typically an equal distribution of heads and tails. Recall that the large number is $n = 1000$ here, which is not in fact very large. It is easy to imagine $n = 10^6$ or even $n = 10^{24}$ (see below), which is the approximate number of molecules in a mole. The table above would "explode" for such high numbers.

If we now ask why we never get, let's say, fewer than hundred heads in a coin-tossing experiment of 1000 tosses, then we find the simple answer just by looking at the table: *because such a sequence would be atypical*, because the overwhelming majority of sequences show heads and tails in almost equal numbers. Perhaps we need a little time to get used to this insight, which comes solely from consideration of large numbers, but getting used to it pays off, because it comes very close to the statistical reasoning of Ludwig Boltzmann.

Here is an example. Consider a lecture hall which is very well insulated when the doors are closed, and suppose the air, consisting of a huge number of molecules (let's say for simplicity 10^{24}), has uniform density in that hall. We now divide the lecture hall mentally into two parts of equal size and ask ourselves: Why does it never happen in practice that all the air molecules occupy the left half of the room? That would be clearly possible, given Newton's laws of motion. To see the answer, think of the following way of encoding this: if the k th molecule is on the left, label it 1, and if it is on the right, label it 0. In this way the configuration of all molecules at a given time is a sequence of ones and zeros, but the length of the sequence is $10^{24} \gg 1000$. Now we are dealing with such enormously large numbers that the above fluctuation from the equilibrium distribution of the molecules is out of the question. Only fluctuations up to about 10^{12} more 0's than 1's (or vice versa), meaning a fraction of about $10^{12}/10^{24} = 10^{-12}$, have any notable chance of appearing.

3.2   Small Causes, Large Effects

Straightforward counting of the kind we just did for coin tossing to justify our everyday experience of randomness actually contains more than meets the eye. The typical sequences also show another important characteristic of chance: the unpredictability of the outcome in each individual coin toss. In other words, in each sequence, the 0's and 1's appear in an irregular manner, but such that the relative frequencies of 0 or 1 will be regular, namely, near $1/2$. So what is the source of this irregular behaviour?

We feel we know the answer intuitively: the coin must be sent spinning and whirling enough to make the result unpredictable, because the smallest change in the initial (angular) momentum imparted to the coin by the throwing hand can lead to a different outcome, from heads to tails or vice versa. The motion of the coin becomes *chaotic*, in the sense that small causes can have large effects. This is very important for the appearance of random behaviour. In fact, it is crucial for the law-like behaviour in random sequences of coin tossings, for instance. The more chaotic the motion, the more stable the regularities in the relative frequencies, and the clearer becomes the law of large numbers. This phenomenon is captured in probability theory by the notion of statistical independence.

But how does this all fit in with the physics, at least when we make a serious attempt to understand it? Naturally, our hand can throw the coin in many different ways, but why does it happen in such a way that independence arises? This question leads to another. The initial conditions for the coin are the initial positions and angular momenta, which are continuous variables, so we can no longer actually count the possibilities. But what can tell us now what is small and what is overwhelmingly large if there is an infinity of possibilities either way? The answer is a measure on the continuum, a *typicality measure*.

3.3   Coarse Graining and Typicality Measures

It is helpful for the discussion of chance to omit human involvement altogether and to think of the coin being tossed, not by a human hand, but by a coin-tossing machine, i.e., a mechanical device, a robot, which takes a coin, tosses it, registers heads or tails, takes the coin again, tosses it again, and so on. This way we know for sure that the situation is pure physics, that is, we have a process that is analogous to the inner workings of a clock. In particular, we now may better appreciate the problem that, with such a machine, there is obviously no place for randomness to begin with.

The situation is not at all an easy one and the aim of this chapter is to help the reader to think this through. The machine must take the coin again and again, let's say for definiteness always with heads up, but must also toss each time with ever so slightly changed initial conditions, in such a way that, with the help of the chaotic motion of the coin in the air, the law of large numbers somehow emerges. The first question we would now like to address is this: How can we capture stochastic independence mathematically for a physical system like our coin-tossing machine?

Let Ω be the physical state space of the machine and coin together. For each $\omega \in \Omega$, the machine produces a coin-tossing sequence, which is completely determined by ω. The result of each coin toss is specified by functions on Ω called *coarse-graining functions*, which map each $\omega \in \Omega$ to the values 0 or 1 (heads or tails). To keep things simple mathematically, we choose as an example $\Omega = [0, 1)$ and $\omega = x \in [0, 1)$, so we are looking for functions which map the interval $[0, 1)$ to the set of values $\{0, 1\}$ and which nevertheless exhibit the characteristic features of the coin toss. One such feature is stochastic independence. The coarse graining must produce this independence in relation with a "natural measure".

The big question when the first rigorous mathematical probability theory was being worked out was: Do such coarse-graining functions exist naturally, i.e., do they arise as natural mathematical objects or must we rely on pathological *ad hoc* constructions? And what should be used as the measure? The answer is at the same time simple and subtle, and it requires a generous helping of modern mathematics.

We represent the initial condition $x \in [0, 1)$ in binary form, i.e., we represent the number in base 2:

$$x = 0. x_1 x_2 x_3 \ldots, \quad x_k \in \{0, 1\}, \quad x = \sum_{k=1}^{\infty} x_k 2^{-k},$$

so that $x_k \in \{0, 1\}$ is the k th binary value of x. Now consider the function r_k which maps x to its k th binary value (see Fig. 3.1 for the graphs for different values of k):

$$r_k : [0, 1) \longrightarrow \{0, 1\}$$

$$r_k(x) = x_k .$$

The functions are called the *Rademacher functions* in honour of Hans Rademacher (1892–1969), who first introduced them. They map each value of $[0, 1)$ to either 0 or 1 and are thus *coarse-graining functions*. The set $r_k^{-1}(\delta)$ is the *pre-image* of $\delta \in \{0, 1\}$, i.e., the set of $x \in [0, 1)$, which are mapped to δ.

We can use the Rademacher functions to model coin tossing. We think of x as representing the physical initial condition and $r_k(x)$ as the final position of the coin in the k th toss, so that r_k is thought of as being given by the solution of the law of motion. The role of the deterministic physical law is thus played by the binary expansion of the numbers.

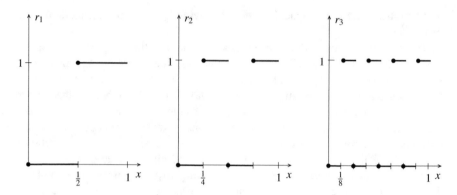

Fig. 3.1 Graphs of the Rademacher functions r_k for $k = 1, 2, 3$. Note that the pre-images are half open intervals

Remark 3.1 In the mathematical literature, coarse-graining functions are called random variables. This notion is somewhat unfortunate, however and was described as "a horrible and misleading terminology" by Kac.[2] This is because the qualification "random" suggests that there is something intrinsically chance-like about them, but there isn't. They are functions, nothing more and nothing less.

Given a value of r_k, what do we learn about x? If $r_1(x) = 1$ then $x \in [1/2, 1)$ and if $r_2(x) = 0$, then $x \in [0, 1/4) \cup [1/2, 3/4)$, etc., but that's all. Suppose we are given the first n values $(x_1, \ldots, x_n)$, then we have more accurate knowledge about x, but we have absolutely no clue about the values x_l for $l > n$. Hence the modelling of the k th toss by r_k seems very fitting: if x is the initial condition and $r_1(x)$ the result of the first toss, $r_2(x)$ that of the second, and so on up to k, then nothing can be concluded for the tosses after k from the first k results.

It gets even better! To see why, it is useful to have some insight into the way the pre-images of the Rademacher functions overlap. For example, the set of initial conditions which lead to the event

in the 1st toss "heads", in the 2nd toss "tails", and in the 4th toss "tails"

is the intersection of the relevant pre-images, i.e.,

$$r_1^{-1}(1) \cap r_2^{-1}(0) \cap r_4^{-1}(0) = [1/2, 9/16) \cup [10/16, 11/16).$$

---

[2]M. Kac, *Statistical Independence in Probability, Analysis and Number Theory.* The Carus Mathematical Monographs, 1959, S. 22.

The set of initial conditions leading to

in the 1st toss "heads" or in the 2nd toss "tails"

is

$$r_1^{-1}(1) \cup r_2^{-1}(0) = [1/2, 3/4).$$

Now we ask whether the Rademacher functions correspond to our idea of statistically independent outcomes in a coin-tossing sequence.

As the name suggests, coarse-graining functions are not one-to-one: many x values are mapped by r_k to one value. As mentioned above, which x values those are is given by the pre-image $r_k^{-1}(\delta)$, $\delta \in \{0, 1\}$, e.g.,

$$r_1^{-1}(0) := \{x \in [0, 1) : r_1(x) = 0\} = [0, 1/2).$$

Coarse-graining functions partition their domain of definition into cells—a characterising feature of such functions. It is exactly this insight which leads us to the weights Laplace attributed to the image values, as discussed at the beginning of Sect. 3.1. The weight $1/2$ attributed to each of the values 0 and 1 is just the *content*, in some sense, of the cells constituting the pre-images. But what is meant by *content*? We have a natural intuition about that, namely, the *content of the interval* $[a, b)$ is its length[3] $\lambda([a, b)) := b - a$, just as the *content of a rectangle* is its area, and then *content* is synonymous with volume in higher-dimensional spaces. In probability theory, the notion of volume has been generalised to that of *measure*, but for intuitive purposes it is always helpful to keep in mind that the mother of all measures is content in the sense just described.

We see from Fig. 3.1, and reflecting for a moment on the graphs for general k, that the content of each pre-image of a value of the Rademacher functions is

$$\lambda\left(r_k^{-1}(\delta)\right) = \lambda(\{x : r_k(x) = \delta\}) = \frac{1}{2}, \quad \delta \in \{0, 1\}.$$

This is the value of the probability $\mathbb{P}$ on the image space, i.e., the value of the probability that we would attribute intuitively. For $\delta \in \{0, 1\}$,

$$\mathbb{P}(\delta \text{ in the } k \text{ th toss}) := \lambda\left(r_k^{-1}(\delta)\right). \tag{3.1}$$

Coarse-graining functions carry the content from the domain of definition to the image space. With this in hand, we now turn to the independence property of Rademacher functions, which corresponds to a key feature of coin tossing.

---

[3] λ is a standard notation for Lebesgue measure, but we do not need the full story here.

For simplicity, we start with the coarse grainings r_1 and r_2. The content has the following product structure:

$$\mathbb{P}(\delta_1 \text{ in the 1st toss and } \delta_2 \text{ in the 2nd toss}) = \lambda\left(r_1^{-1}(\delta_1) \cap r_2^{-1}(\delta_2)\right)$$

$$= \lambda(\{x : r_1(x) = \delta_1 \cap r_2(x) = \delta_2\})$$

$$= 1/4 = (1/2)^2$$

$$= \lambda\left(r_1^{-1}(\delta_1)\right)\lambda\left(r_2^{-1}(\delta_2)\right). \qquad (3.2)$$

This product structure is taken to define statistical independence, i.e., the Rademacher functions r_1 and r_2 are indeed statistically independent.

Actually, this shouldn't come as too much of a surprise. It was to be expected because the Rademacher functions map to values in the binary expansion and, as we noted earlier, the first binary digit has no implications for the second binary digit, which is indeed the naive notion of independence. On the other hand, if the above result about the product structure seems too trivial to be of any great importance, then think again! There is a crucial point to be grasped here.

Remark 3.2 We could have used the word *probability* and we could have said that the probability that r_1 takes the value δ_1 *and* r_2 takes the value δ_2 equals the product of the individual probabilities. But what would that have done for us? The product structure for the content of special sets is a mathematical fact, while probability always carries a subjective undertone, which we have no use for here.

For a general collection of Rademacher functions, letting $n_k \in \mathbb{N}$, $k = 1, 2, \ldots, n$, $\delta_{n_k} \in \{0, 1\}$, we have

$$\lambda\left(\bigcap_{k=1}^{n} r_{n_k}^{-1}(\delta_{n_k})\right) = \lambda\left(\bigcap_{k=1}^{n} \{x : r_{n_k}(x) = \delta_{n_k}\}\right)$$

$$= (1/2)^n = \prod_{k=1}^{n} \lambda\left(r_{n_k}^{-1}(\delta_{n_k})\right). \qquad (3.3)$$

That gives us the necessary trust that *statistical independence* is not just a notion invented by humans to satisfy their intuition about the way a coin-tossing experiment should be described, but rather it can be a natural mathematical property of the contents of the cells in the partition induced by coarse-graining functions. Therefore the product structure can be accepted as a good mathematical definition of statistical independence. What we have done so far then is to see a way to grasp randomness in physics without introducing any kind of *random* elements.

The independence of the Rademacher functions can be seen "directly". The coarse graining yields a very distinct partition. Consider again Fig. 3.1 and recall how the pre-images of r_k for large k partition the interval $[0, 1)$ further. In particular, note how the pre-images of different coarse grainings mix in a precisely coordinated way, as expressed in (3.3). That mixing is the way independence should be thought of—at least in principle: the very orderly way in which the pre-images of the Rademacher functions intertwine is the ideal case, while in realistic physical models the mixing will be much less perfect and orderly, and harder to picture.

The fact that realistic mixing may look very different is also due to another factor. Indeed, the mixing is not only perfect because the Rademacher functions partition the interval the way they do, but also because we measure the content of the cells by the length of the interval (the Lebesgue measure). This is the most natural measure (since it is unbiased), but it is nevertheless particular among all possible measures. The abstract notion of measure generalizing our a priori intuition of the content (volume) of a set—which arises in physics as well and which we shall also use later—may weight the cells of a partition in what we would intuitively call a non-uniform way. At the end of the day, it is the physical law which decides the measure.

To sum up, what we should bear in mind here is that the degree of statistical independence is determined by two things: (1) the kind of coarse-graining function which maps the fundamental set to the set of relevant events or outcomes we are focusing on and (2) the measure which determines the "size" of the pre-image cells.

Here is an example of another possible (actually arbitrarily chosen) measure. Instead of λ, take $\mu = 4\lambda/3$ on $[0, 3/4)$ and $\mu = 0$ on $[3/4, 1)$. Then $\mu([0, 3/4)) = 1$ and

$$\mu([1/2, 1)) = \frac{4}{3}\lambda([1/2, 3/4)) = \frac{4}{3}\left(\frac{3}{4} - \frac{1}{2}\right) = \frac{1}{3}.$$

If we use this measure to determine the size of the cells in the partition effected by the Rademacher functions, we find

$$\mu\left(r_1^{-1}(0) \cap r_2^{-1}(1)\right) = \mu(\{x : r_1(x) = 0\} \cap \{x : r_2(x) = 1\})$$

$$= \mu([1/4, 1/2)) = \frac{4}{3}\lambda((1/4, 1/2]) = \frac{4}{3} \times \frac{1}{4} = \frac{1}{3}$$

$$\neq \mu\left(r_1^{-1}(0)\right)\mu\left(r_2^{-1}(1)\right)$$

$$= \frac{4}{3}\lambda([0, 1/2))\frac{4}{3}\lambda([1/4, 1/2)) = \frac{2}{9}.$$

The product structure has gone, and hence also the independence.

The notion of *independence* of random variables or random events is defined in every textbook on probability without much ado. We have performed a song and dance about it here to stress the fact that the mathematical foundation is based on the particular way in which the pre-images of the coarse-graining functions mix and also on the measure used to determine the size of the cells. We did this to get a deeper understanding of what is going on. The Rademacher functions are the *archetypal independent random variable* and it was only after their discovery that probability theory was developed into a serious mathematical discipline.

3.4   The Law of Large Numbers

Table 3.1 tells us that *the relative frequency of heads and tails is typically around 1/2*. The law of large numbers can thus be read directly off the table. So where is the chaos? Where is the independence? The answer is that it is in the counting of sequences! This is something we need to think hard about! How do we get to the typicality statement from the fundamental level, for example, by modelling coin tossing using the Rademacher functions? We can no longer count discrete entities on that fundamental level, since it is a continuum. The measure λ now takes over the role of counting. Instead of talking about the overwhelming majority in terms of counting, we must now define the overwhelming majority in terms of our measure. To come to grips with that, we have to ask how we can formulate the relative frequency of 0's and 1's (heads and tails), which is the thing we are interested in. In fact, this is done using something called the *empirical distribution*. Introducing the indicator function $\mathbb{1}_A(x)$ of a set A, which is 1 if $x \in A$ and 0 otherwise, we have the following definition:

Definition 3.1   The function

$$\rho_{\text{emp}}^n(\{\delta\}, x) := \frac{1}{n} \sum_{k=1}^{n} \mathbb{1}_{\{\delta\}}(r_k(x)), \qquad \delta \in \{0, 1\}, \quad x \in [0, 1),$$

is called the *empirical distribution* of the Rademacher functions.

It is a coarse-graining function (a random variable), coarse graining $[0, 1)$, and it determines the relative frequency of the appearance of the digit δ in the first n binary digits of x.

Remark 3.3   The word "empirical" could be confusing because, on the one hand, there is the empirical distribution which is given by the relative frequencies determined in an experiment, and on the other, there is the theoretical entity above for which the physical theory will produce the prediction for the experimental outcome. But "theoretical empirical distribution" is a bit too long and uses adjectives which seem even to contradict each other. In short, $\rho_{\text{emp}}^n(\{\delta\}, x)$ captures the *empirical* distribution of the coin-tossing results *within* our theory for the initial condition x. We shall soon get serious about this, in fact in Theorem 3.3.

We shall now show that typically, for large n, the relative frequency for $\delta = 1$ is about $1/2$, and likewise for $\delta = 0$. This means that the coarse-graining function $\rho_{\text{emp}}^n(\{\delta\}, x)$ partitions the interval $[0, 1)$ into cells of different sizes, where the cell which is mapped to values near $1/2$ has almost size 1. There are many smaller cells on which the function takes values away from $1/2$, i.e., other values are possible but don't have enough weight to be taken seriously when judged by the content or size of the cells.

We wish to establish this mathematically. To do so, consider the deviation

$$\left| \rho_{emp}^n(\{\delta\}, x) - \frac{1}{2} \right| > \epsilon \,,$$

for some $\epsilon > 0$ (thinking of ϵ as a small number), and compute the corresponding cell size. This will lead us to the *law of large numbers* which is also often called *the law of the mean*. Hence we estimate the content of the set

$$\left\{ x \in [0, 1) : \left| \rho_{emp}^n(\{\delta\}, x) - \frac{1}{2} \right| > \epsilon \right\}, \quad \epsilon > 0 \,.$$

To do this, we write the content of a set in terms of an integral, using the indicator function of the set. In general terms,

$$\lambda\big(\{x : |f(x)| > \epsilon\}\big) = \int_0^1 \mathbb{1}_{\{x:|f(x)|>\epsilon\}}(x)\mathrm{d}x \,. \tag{3.4}$$

Next observe that, if $|f(x)| > \epsilon$, then $(|f(x)|/\epsilon)^n > 1$, for all $n \in \mathbb{N}$. We then have the following very effective inequality (although simple, it deserves a moment's reflection):

$$\mathbb{1}_{\{x:|f(x)|>\epsilon\}}(x) \le \left(\frac{|f(x)|}{\epsilon} \right)^n, \quad \forall n \in \mathbb{N} \,.$$

Putting this into (3.4), we get

$$\lambda\left(\{x : |f(x)| > \epsilon\}\right) = \int_0^1 \mathbb{1}_{\{z:|f(z)|>\epsilon\}}(x)\,\mathrm{d}x \le \int_0^1 \left(\frac{|f(x)|}{\epsilon} \right)^n \mathrm{d}x \,. \tag{3.5}$$

For $n = 2$, this is known as the *Chebyshev inequality*. For general n, it is called the *Markov inequality*. We shall use it now for $f(x) = \rho_{emp}^n(\{\delta\}, x) - 1/2$. To save work, we note that, in the context of the Rademacher functions, $\mathbb{1}_{\{1\}}(r_k(x)) = r_k(x)$, so the notation simplifies in the case $\delta = 1$, yielding

$$\rho_{emp}^n(\{1\}, x) - \frac{1}{2} = \frac{1}{n} \sum_{k=1}^n \left[r_k(x) - \frac{1}{2} \right] \,.$$

Thus,

$$\lambda\left(\left\{x : \left|\rho_{emp}^n(\{1\}, x) - \frac{1}{2}\right| > \epsilon\right\}\right) \leq \frac{1}{\epsilon^2}\int_0^1 \left(\rho_{emp}^n(\{1\}, x) - \frac{1}{2}\right)^2 dx$$

$$= \frac{1}{n^2\epsilon^2}\int_0^1 \left[\sum_{k=1}^n \left(r_k(x) - \frac{1}{2}\right)\right]^2 dx.$$

Expanding the square yields a diagonal sum over n terms and an off-diagonal sum with $n(n-1) \approx n^2$ terms. Setting $a_k = r_k - 1/2$, we obtain

$$\left(\sum_{k=1}^n a_k\right)^2 = \sum_{k=1}^n a_k^2 + \sum_{k \neq l=1}^n a_k a_l.$$

This is where we can use independence. To integrate the off-diagonal terms, we use the fact that, for *independent* coarse-graining functions, the integral of a product equals the product of the integrals:

$$\int_0^1 \left[r_k(x) - \frac{1}{2}\right]\left[r_l(x) - \frac{1}{2}\right] dx = \int_0^1 \left[r_k(x) - \frac{1}{2}\right] dx \int_0^1 \left[r_l(x) - \frac{1}{2}\right] dx = 0.$$

$$(3.6)$$

To see the last equality, recall the graphs in Fig. 3.1 and note that each integral is in fact zero. We are then left with the diagonal terms, but there is nothing to compute here either, because $(r_k - 1/2)^2 = 1/4$ for all k, so

$$\sum_{k=1}^n \int_0^1 \left(r_k(x) - \frac{1}{2}\right)^2 dx = \frac{n}{4}.$$

We have thus proved the following theorem:

Theorem 3.1 (Law of Large Numbers for Rademacher Functions)   *For all* $\epsilon > 0$,

$$\lambda\left(\left\{x \in [0, 1) : \left|\rho_{emp}^n(\{\delta\}, x) - \frac{1}{2}\right| > \epsilon\right\}\right) \leq \frac{1}{4n\epsilon^2}, \quad \delta \in \{0, 1\}.\qquad (3.7)$$

Note as an aside that the "expected value" of ρ_{emp}^n *is*

$$\int_0^1 \rho_{emp}^n(\{\delta\}, x) dx = \frac{1}{2}.$$

We emphasize that the right-hand side gets arbitrarily small with increasing n, hence the reference to large numbers. The theorem can be restated in different but equivalent ways:

- The set of $x \in [0, 1)$ for which the relative frequency of 1's and 0's in the binary expansion is not close to 1/2 has negligible content.
- For the *overwhelming* majority of $x \in [0, 1)$, the relative frequencies of the binary digits are close to 1/2.
- It is *typical* that the relative frequencies of 1's and 0's are approximately 1/2.
- Using the Rademacher functions as a model for physical coin-tossing experiments, heads and tails typically appear more or less equally often. We thus make the following prediction: a coin-tossing experiment will reveal heads and tails more or less equally often.
- We may call the typical value 1/2 of ρ_{emp}^n (its expected value) the *probability* for heads or tails, which is the Laplace probability for coin tossing.

We thus learn that the Laplace probability, which is essentially based on counting, can be explained by a coarse-graining function on a continuum, which is closer to the physics of the coin-tossing process. We shall focus more closely on this in the next section.

3.5 Typicality in the Continuum

We first talked about typicality in terms of counting. That is, we counted those sequences with a relevant feature (relative frequencies of "heads" and "tails" approximately equal to 1/2) and found that they made up the overwhelming majority of possible sequences. But the molecules in the example of the lecture hall are moving around and their motion is determined by specifying the position and velocity of all the molecules in the hall. Likewise with coin tossing, the conditions which determine the motion are to be drawn from a continuum. This remains hidden when we do the counting. In other words, the counting is applied on a coarse-grained level of description. The question is thus: How can we extend the idea of typicality to the continuum?

We already opened the way in the last section. The counting is to be replaced by a *measure*, and in particular, by a measure of typicality. The measure assigns weights to subsets of the physically relevant space, i.e., to subsets of the *phase space*, as Boltzmann called it. The important question is this: What or who determines the typicality measure? Since the typicality measure determines what is typical and what is not, or in other words what it is that we actually experience (because we experience what is typical) among all possibilities, the typicality measure should have a clearly objective character. Put another way, it is simply a physical fact (independent of what we humans think or know) that the lecture hall is homogeneously filled with air molecules, so that no student in the right half of the lecture hall will suddenly suffocate.

Therefore we investigate the question in the context of the most famous deterministic physical theory, namely Newtonian mechanics, and explain how that theory itself determines the typicality measure. But what we shall say applies (in an appropriate form) to all physical theories. Any physical theory will in some way or other be given by a *dynamical system* $(\Omega, \mathbb{P}, T_t)$. The first ingredient is a phase space Ω, for example, the set of all possible positions and velocities of all gas molecules in the lecture hall. In addition, there is a time evolution $T_t : \Omega \mapsto \Omega$, where $t \in \mathbb{R}$ stands for time, for example, the time evolution of the positions and velocities of all relevant particles. Finally, there is a measure $\mathbb{P}$ which remains unchanged under the time evolution. This means that $\mathbb{P}$ is *stationary* with respect to T_t.

To grasp the meaning of stationarity, we must first consider how a general measure $\mathbb{P}$ changes with the time evolution T_t on Ω. In fact, for all (measurable) subsets $A \subset \Omega$,

$$\mathbb{P}_t(A) := \mathbb{P}(T_t^{-1}(A)) \,.$$

This just states the obvious: the measure at time t of a set A is nothing but the measure of that set which evolved over time t into A, viz.,

$$T_t^{-1}(A) = \{\omega | T_t \omega \in A\} \,.$$

Now we can ask whether there exists a *distinguished* measure which does not change, i.e., a *stationary* measure with the property $\mathbb{P}_t(A) = \mathbb{P}(A)$. This would be so if and only if $\mathbb{P}(A) = \mathbb{P}(T_t^{-1}(A))$ for all subsets A. If taken as the typicality measure, the stationary measure yields the nice property that typicality is timeless. That in turn means that statistical reasoning based on typicality is possible at any time! To stress its importance we repeat here: the measure $\mathbb{P}$ is stationary with respect to the time evolution T_t if, for all A and t,

$$\mathbb{P}_t(A) := \mathbb{P}(T_t^{-1}(A)) = \mathbb{P}(A) \,. \tag{3.8}$$

The next question is a practical one: How can we determine the stationary measure for a given physical theory, such as Newtonian mechanics?

3.5.1 Newtonian Mechanics in Hamiltonian Form

The time evolution T_t in Newtonian mechanics describes the time evolution of point particles. The Newtonian theory is of second order which means that the dynamical law is given by a second order differential equation for the positions of the particles $q_i \in \mathbb{R}^3$, viz.,

$$m_i \frac{d^2 \mathbf{q}_i}{dt^2} = \mathbf{K}_i(\mathbf{q}_1, \dots, \mathbf{q}_N) \,, \quad i = 1, \dots, N \,, \tag{3.9}$$

where the m_i are parameters, called masses, and K is a force field, e.g., the gravitational force given by

$$\mathbf{K}_i(\mathbf{q}_1, \ldots, \mathbf{q}_N) = \sum_{j \neq i} Gm_i m_j \frac{\mathbf{q}_j - \mathbf{q}_i}{|\mathbf{q}_j - \mathbf{q}_i|^3},$$

with G the gravitational constant.

The phase space of Newtonian mechanics for an N-particle system is $\mathbb{R}^{6N}$, because the elements needed to specify the time evolution T_t of the system are all the positions and velocities of all the particles (i.e., three position coordinates and three velocity components per particle). This is because the Newtonian law is a second order differential equation. A more prosaic representation of the law is achieved by reduction of the second order differential equation to a first order one, directly defined on phase space $\mathbb{R}^{6N}$. For notational convenience, we introduce

$$(q, p) := (\mathbf{q}_1, \ldots, \mathbf{q}_N, \mathbf{p}_1, \ldots, \mathbf{p}_N), \quad \text{with momentum } \mathbf{p}_i = m_i \dot{\mathbf{q}}_i := m_i \frac{d\mathbf{q}_i}{dt},$$

and call $\Omega = \mathbb{R}^{3N} \times \mathbb{R}^{3N}$ the *phase space*[4] of elements $\omega = (q, p)$. Equation (3.9) becomes

$$\begin{aligned}
\dot{q} &= m^{-1} p, \\
\dot{p} &= K(q) = \left(\mathbf{K}_1(\mathbf{q}_1, \ldots, \mathbf{q}_N), \ldots, \mathbf{K}_N(\mathbf{q}_1, \ldots, \mathbf{q}_N)\right),
\end{aligned} \tag{3.10}$$

where m is the mass matrix, a diagonal matrix with m_i as entries.

In the case of gravitation, and in many other cases as well, it turns out that K is a gradient of a potential, i.e., $K = -\nabla V$. Using this, we can write (3.10) as

$$\begin{aligned}
\dot{q} &= \frac{\partial H}{\partial p}(q, p), \\
\dot{p} &= -\frac{\partial H}{\partial q}(q, p),
\end{aligned} \tag{3.11}$$

where the so-called Hamiltonian function[5] is given by

$$H(q, p) = \frac{1}{2} \langle p, m^{-1} p \rangle + V(q) := \frac{1}{2} \sum_{i=1}^{N} \frac{\mathbf{p}_i^2}{m_i} + V(\mathbf{q}_1, \ldots, \mathbf{q}_N). \tag{3.12}$$

---

[4]In classical physics, the symbol Γ is used for phase space, following Boltzmann. In the mathematical foundations of probability theory, Andrey Nikolaevich Kolmogorov (1903–1987) introduced instead the symbol Ω, which may be more suggestive of the "fundamental set".

[5]Introduced by William Rowan Hamilton (1805–1865) and denoted by H in honour of Christiaan Huygens (1629–1695).

It is helpful to think of *Hamiltonian mechanics*, as given by (3.11) and (3.12), as an independent physical theory, although in the case considered here it is equivalent to Newtonian mechanics.

Hamiltonian mechanics is in a certain sense unromantic, in contrast to the more intuitive Newtonian mechanics (see below). If one considers the Hamilton function as representative of the fundamental law of motion, the mechanics goes like this. The Hamilton function of an N-particle system $H(q, p)$ generates the following vector field on phase space Ω:

$$
v^H(q, p) = \begin{pmatrix} \dfrac{\partial H}{\partial p}(q, p) \\[2ex] -\dfrac{\partial H}{\partial q}(q, p) \end{pmatrix} = \begin{pmatrix} \dfrac{\partial H}{\partial \mathbf{p}_1}(q, p) \\ \vdots \\ \dfrac{\partial H}{\partial \mathbf{p}_N}(q, p) \\ -\dfrac{\partial H}{\partial \mathbf{q}_1}(q, p) \\ \vdots \\ -\dfrac{\partial H}{\partial \mathbf{q}_N}(q, p) \end{pmatrix} .
$$

Here $\partial H/\partial \mathbf{q}_k$ and $\partial H/\partial \mathbf{p}_k$ are the gradients with respect to the vectors $\mathbf{q}_k$ and $\mathbf{p}_k$, that is, differentiating with respect to each of the three coordinates in each case. Equations (3.11) tell us that the integral curves along the vector field (i.e., such that the vector field is tangent everywhere to these integral curves) are the curves giving the temporal evolution of the possible states

$$
(Q(t), P(t)) = (\mathbf{Q}_1(t), \dots, \mathbf{Q}_N(t), \mathbf{P}_1(t), \dots, \mathbf{P}_N(t)) , \quad t \in \mathbb{R} ,
$$

of the system, where the tuple $(\mathbf{Q}_k(t), \mathbf{P}_k(t))$ indicates the position and the velocity (or rather, the momentum) of the kth particle at time t. This yields a picture made up of curves, namely the curves showing the evolution of the system as time goes by, in a high-dimensional space. The picture is actually rather hard to depict! Recall our discussion of the configuration space and note that phase space is configuration space times "momentum space". Applying this picture to the gas molecules in the lecture hall, the dimension is about 10^{24}.

So why did we say the picture was unromantic? Newtonian mechanics talks about masses which attract each other or masses acting upon each other via forces—an almost human behaviour. That all disappears in the Hamiltonian picture. There is only a mathematical function, the Hamilton function, which generates a vector field on phase space. The integral curves of the vector field define a flow, the *Hamiltonian flow* $(T_t^H)_{t \in \mathbb{R}}$ on Ω which maps the "initial" values (q, p) to the time-evolved values at time t along the trajectories:

$$
T_t^H : \Omega \to \Omega, \quad (q, p) \mapsto \left(q\big(t, (q, p)\big), p\big(t, (q, p)\big) \right), \quad t \in \mathbb{R} ,
$$

where $q\left(0, (q, p)\right) = q$, $p\left(0, (q, p)\right) = p$ and

$$\frac{\mathrm{d}T_t^H(\omega)}{\mathrm{d}t} = v^H\left(T_t^H(\omega)\right), \quad \omega = (q, p). \tag{3.13}$$

To turn Hamiltonian mechanics into a dynamical system, we need a stationary measure $\mathbb{P}$. Our next task will be to find that.

3.5.2 Continuity Equation and Typicality Measure

For notational simplicity we suppress the index H on T_t and consider (3.8) in the slightly more general form

$$\int f\left(T_t(\omega)\right)\mathrm{d}\mathbb{P}(\omega) = \int f(\omega)\mathrm{d}\mathbb{P}(\omega). \tag{3.14}$$

For $f = \mathbb{1}_A$, we get back

$$\mathbb{P}_t(A) := \mathbb{P}\left(T_{-t}(A)\right) = \mathbb{P}(A).$$

Note in passing that the Hamiltonian flow is invertible:

$$T_{-t}T_t = \mathrm{id}.$$

The problem now is to use (3.14) to actually find such a $\mathbb{P}$. The trick is to turn (3.14) into a differential equation. For that we need to assume that $\mathbb{P}$ has a density, i.e., $\mathbb{P}(\mathrm{d}\omega) = \rho(\omega)\mathrm{d}\omega$. That is plausible since Ω is a continuum. So we rewrite (3.14) in the form

$$\int f(\omega)\rho(\omega, t)\mathrm{d}\omega := \int f\left(T_t(\omega)\right)\rho(\omega)\mathrm{d}\omega = \int f(\omega)\rho(\omega)\mathrm{d}\omega, \tag{3.15}$$

where we now think of f as being a test function, i.e., differentiable and zero at infinity. Note that on the left we have defined the *time dependent density* $\rho(\omega, t)$. It is the density of $\mathbb{P}_t$, the measure transported by T_t. If we wish to picture that, we recall that each flow line of the Hamiltonian flow on Ω represents a trajectory of the system, i.e., the temporal evolution of the entire N-particle system. We view the measure as assigning *continuous* weights to the system trajectories and, due to the way the trajectories evolve, separating from each other or getting closer in time,[6] the distribution of weights changes with time. This is expressed by $\rho(\omega, t)$.

---

[6]Flow lines (integral curves) can never cross each other since they are defined by a vector field. If they did, then at a crossing point, there would be two tangential "velocity" vectors, which is impossible for a vector field unless the vector field is zero at that point.

The density $\rho(\omega, t)$ obeys a differential equation, in fact a transport equation called the *continuity equation*. To see that, we differentiate the left-hand side and the middle of (3.15) with respect to t. Using the chain rule, this yields

$$\int f(\omega)\frac{\partial \rho(\omega, t)}{\partial t}d\omega = \int \frac{dT_t(\omega)}{dt} \cdot \nabla f\left(T_t(\omega)\right) \rho(\omega)d\omega, \qquad (3.16)$$

where the dot signifies the scalar product in $\mathbb{R}^{6N}$. Since by (3.13)

$$\frac{dT_t(\omega)}{dt} = v\left(T_t(\omega)\right),$$

the right-hand side of (3.16) becomes

$$\int \rho(\omega)v\left(T_t(\omega)\right) \cdot \nabla f\left(T_t(\omega)\right) d\omega.$$

By the definition (3.15) of $\rho(\omega, t)$, this is equal to

$$\int \rho(\omega, t)v(\omega) \cdot \nabla f(\omega) d\omega.$$

Assuming that f decays quickly to zero towards infinity, partial integration then turns this into

$$-\int f(\omega)\mathrm{div}(v(\omega)\rho(\omega, t)) d\omega.$$

Thus (3.16) becomes

$$\int f(\omega)\frac{\partial \rho(\omega, t)}{\partial t}d\omega = -\int f(\omega)\mathrm{div}(v(\omega)\rho(\omega, t))d\omega.$$

Since f is arbitrary, we can read off the continuity equation:

$$\frac{\partial \rho(\omega, t)}{\partial t} = -\mathrm{div}\left(v(\omega)\rho(\omega, t)\right). \qquad (3.17)$$

Remark 3.4 The equation means that there is no loss of "mass" (in the sense of weight). Integrating (3.17) over a volume $V \subset \Omega$ and using Gauss' theorem yields

$$\frac{\partial}{\partial t}\int_V \rho(\omega, t)d\omega = -\int_{\partial V} \rho(\omega, t)v(\omega) \cdot d\sigma,$$

where the surface integral of the *flux* through the surface appears on the right-hand side. Any change of "mass" in the phase volume V can only occur by mass flowing out or coming in through the boundary ∂V of V. Equation (3.17) can be written more economically in the form

$$\frac{\partial \rho}{\partial t} + \operatorname{div} J = 0 \,,$$

where $J := \rho v$ is the "flux" in phase space.

Now let us return to the question of stationarity as expressed in (3.15). We look for a stationary (time independent) solution $\rho(\omega, t) = \rho(\omega)$ of the continuity equation (3.17) for the Hamiltonian vector field v^H. Using the product rule, we get

$$\frac{\partial \rho(\omega, t)}{\partial t} = -\operatorname{div}(v^H(\omega)\rho(\omega, t)) = -\rho(\omega, t)\operatorname{div}(v^H(\omega)) - \operatorname{grad} \rho(\omega, t)v^H(\omega) \,,$$

where $\operatorname{div}\big(v^H(\omega)\big)$ drops out because

$$\operatorname{div} v^H = \left(\frac{\partial}{\partial q}, \frac{\partial}{\partial p}\right) v^H = \begin{pmatrix} \frac{\partial}{\partial q} \\ \frac{\partial}{\partial p} \end{pmatrix} \cdot \begin{pmatrix} \frac{\partial H}{\partial p} \\ -\frac{\partial H}{\partial q} \end{pmatrix} = \frac{\partial^2 H}{\partial q \partial p} - \frac{\partial^2 H}{\partial p \partial q} = 0 \,. \quad (3.18)$$

This last result is known as *Liouville's theorem*.

Theorem 3.2 (Liouville's Theorem)

$$\operatorname{div} v^H = 0 \,.$$

We immediately obtain a stationary solution of the continuity equation for the Hamiltonian flow, i.e., with $\partial \rho(\omega, t)/\partial t = 0$, because the equation which remains when Theorem 3.2 is taken into account is

$$0 = -\operatorname{grad} \rho(\omega, t) \, v^H(\omega) = -\nabla \rho(\omega, t) \cdot v^H(\omega) \,. \quad (3.19)$$

That means that the Lebesgue measure $d\omega = d^{3N}q \, d^{3N}p$ on the N-particle phase space is a stationary measure. Isn't that amazing! The first physical theory of interacting particles, i.e., Newtonian mechanics, yields the most natural measure— the volume—as stationary measure. A common way of saying this is that the Hamiltonian flow is *volume preserving*.

Before we discuss this further we ask whether there are other stationary measures. Indeed, there are, as we shall show now. On the right-hand side of (3.19), we have

$$v^H \cdot \nabla \rho = \left(\frac{\partial H}{\partial p} \cdot \frac{\partial}{\partial q} - \frac{\partial H}{\partial q} \cdot \frac{\partial}{\partial p} \right) \rho$$

$$= \left(\dot{q} \cdot \frac{\partial}{\partial q} + \dot{p} \cdot \frac{\partial}{\partial p} \right) \rho(q, p)$$

$$= \frac{\mathrm{d}}{\mathrm{d}t} \rho\big(q(t), p(t)\big).$$

The last expression is the change in the function ρ along the system trajectories. A stationary measure is then one whose density is constant along the trajectories. It is straightforward to show that one such function is $H(q, p)$, a property usually referred to as conservation of energy. Hence every function $f(H(q, p))$ stays constant along trajectories.

One choice often used in statistical physics is the *canonical distribution*, a density which, by integration with respect to $\mathrm{d}^{3N}q \, \mathrm{d}^{3N}p$, gives rise to the *canonical measure* or *canonical ensemble*

$$\rho_\beta = f(H) = \frac{\mathrm{e}^{-\beta H}}{Z(\beta)}, \tag{3.20}$$

where β is interpreted thermodynamically as $\beta = 1/k_B T$, with k_B the Boltzmann constant and T the temperature. $Z(\beta)$ is a normalisation factor.

Going back to the volume measure, energy conservation partitions the phase space Ω into shells of constant energy Ω_E:

$$\Omega_E = \{(q, p) : H(q, p) = E\}, \quad \Omega = \bigcup_E \Omega_E.$$

Therefore if we think of an isolated system, i.e., one which has no exchange of any kind with its environment, then the system will remain on one of the shells Ω_E during its time evolution. Hence, the density

$$\rho_E = \frac{1}{Z} \delta\big(H(q, p) - E\big) \tag{3.21}$$

is also stationary (Z is the normalisation constant). When multiplied by the volume element $\mathrm{d}^{3N}q \mathrm{d}^{3N}p$, this becomes the "content" measure $\mathbb{P}_E$ on the energy shell Ω_E. It is variously referred to as the *microcanonical measure*, the *microcanonical ensemble*, or the *microcanonical distribution*. It is more fundamental than the canonical measure, because the latter can be obtained as the *typical* distribution of a subsystem of a large system with the microcanonical measure as typicality measure. What we mean by "typical distribution" and "is distributed according to" will be explained next.

3.5.3 Typicality and the Statistical Hypothesis

In statistical physics, we use statements like "the system is canonically distributed", or again "$\mathbb{P}_E$ is the probability measure on the system's energy surface Ω_E". What we are talking about here is a statistical description of *ensembles* of "identical" subsystems of our universe. This could be an ensemble of spatially separated but otherwise identical subsystems (many different people each toss a coin) or an ensemble which refers to the same subsystem but at different times (one coin is tossed many times). When we talk about a distribution, we are referring to a *statistical hypothesis* about the relative frequencies of particular variables of the system across the ensemble (like the relative frequencies of heads or tails). What justifies the statistical hypothesis? The careful reader will note that we have already asked this question, and indeed all the foregoing has revolved around it.

This question is related to another, which concerns the applicability of the notion of probability to larger and larger systems, eventually to the largest conceivable system, namely, the universe itself. To see this, note first that any system is in general a subsystem of a larger system with which it may interact, or not, if the subsystem is sufficiently isolated from its environment. But independently of the size of the system, i.e., the number of particles it comprises, it is always the same physical law which applies, namely Hamiltonian mechanics with an appropriate Hamiltonian function. The Hamiltonian formulation of mechanics does not involve any limit on the number of particles, nor on the spatial size of the system. This is true for all fundamental theories. They are not stamped with an expiry date, or any indication of size which must not be exceeded for the theory to remain valid.[7] In fact, the fundamental theory is always a theory about the biggest conceivable system, viz., the universe.

Does it make sense to formulate a statistical hypothesis for the universe in order to understand the measure $\mathbb{P} = \mathbb{P}_{H_U}$ which is stationary with respect to the Hamiltonian flow generated by H_U, the Hamiltonian function of the universe? In fact, it does not. What could we do with an empirical distribution across an ensemble of universes? We live in and have access to only one universe—our own. So does that mean that a measure like $\mathbb{P} = \mathbb{P}_{H_U}$ on the phase space of the universe has no meaning at all? In fact, it does have a meaning, and it is absolutely necessary. It is a typicality measure, but should be strictly distinguished from any vague notion of probability, with its various interpretations, such as relative frequencies or degree of

---

[7]That does not mean that a theory which is considered fundamental at the present time cannot be superseded by another one. For example, classical physics gets incorporated into quantum physics, and the realm of validity of the old theory is understood in the context of the new fundamental theory (see Sect. 1.6).

belief. The latter is harmless when applied to subsystems and the former is needed to justify the application. In short, a *probability measure* describes typical empirical distributions of an ensemble of subsystems. The *typicality measure* on the larger encompassing system (like the universe) defines what typical means.

Two questions remain open. First, how can we guess a sensible statistical hypothesis for let's say the N-particle system—as a subsystem of the universe—with Hamiltonian function H? A natural choice (for an equilibrium situation) is one which is given by a stationary measure. Various possibilities recommend themselves depending on the physical situation. If the system is sufficiently isolated, the so-called *microcanonical ensemble* (3.21) might be appropriate. The statistical hypothesis can then be expressed as follows:

Statistical Hypothesis In an ensemble of similar isolated subsystems which all have energy E, the empirical distribution of the coordinates (q, p) is typically given by $\mathbb{P}_E$.

The subsystem may not be completely isolated but engage in some (weak) interaction with its (large) environment by exchanging energy (heat). Suppose that for the large environment the microcanonical measure is applicable as typicality measure. Then we find that the *canonical ensemble* (3.20) is a suitable statistical hypothesis for the subsystem. A simple special case is an ideal gas, with energy E, particle number N, and temperature

$$T = \frac{2}{3} \frac{E}{N k_{\mathrm{B}}} .$$

We can consider each individual particle as a subsystem with Hamiltonian function

$$H = \frac{1}{2} \frac{\mathbf{p}^2}{m} ,$$

whereupon (3.20) is then the Maxwellian velocity distribution. We may then make the assertion:

For typical configurations of the ideal gas with temperature T, i.e., in thermal equilibrium, the momenta of the particles are approximately distributed like

$$\rho_{\mathrm{Max}}(\mathbf{p}) = Z^{-1} \mathrm{e}^{-\mathbf{p}^2 / 2 m k_{\mathrm{B}} T} .$$

In this case, the large system is the gas system as a whole, with typicality measure $\mathbb{P}_E$, and the ensemble consists of the $1 \ll M \ll N$ particles whose statistics we are attempting to describe.

The subsystem may also exchange particles with the environment so that the number of particles in the subsystem fluctuates, but with an average constant density λ of particles. The corresponding distribution is called the *grand canonical ensemble*. For the ideal gas, it is a Poisson distribution on the phase space of the ideal gas with density $\rho_{\mathrm{Max}}(\mathbf{p})\lambda$.

The statistical hypothesis seems like a more or less natural ansatz for making statistical propositions in physics. But a hypothesis is a hypothesis no matter what, and therefore it must be justified, or even better, it must be proven. And that brings us to the second question: How can we prove a statistical hypothesis? The answer is that we must establish the law of large numbers for empirical distributions. This in turn means that we must show that, in a *typical* universe, i.e., in the overwhelming majority of possible universes, the (theoretical) empirical distribution of the coordinates (q, p) in an ensemble of (small) subsystems is given approximately by the distribution corresponding to the statistical hypothesis. This leads us to another question: typical with respect to which measure?

A natural idea, due to Boltzmann, is that the physical law itself should suggest the typicality measure. How? *By requiring the measure to be stationary.*[8] Stationarity thus appears in a new light. The flow on phase space is an expression of the physical law and the stationarity requirement yields a distinguished measure. If we wished, we could add to the dynamical law the law that the statistical analysis must be carried out using the distinguished measure.

To justify the statistical hypothesis we need to take the notion of subsystem, or rather the notion of an ensemble of subsystems, a little more seriously. We thus note that, if ω represents the phase space point of the actual universe, then it contains all the phase space coordinates of the subsystems which form the ensemble, i.e., we can write $\omega = (\omega_E, \omega_R)$, where ω_E contains the phase space points of all subsystems of the ensemble and ω_R contains the phase space coordinates of the environment of the ensemble, i.e., of all the "rest". The N-particle system, i.e., each of the M similar members of the ensemble, has the phase space coordinates ω_{S_i}, $i = 1, \ldots, M$, contained in ω_E. Each subsystem in the ensemble is described by the Hamiltonian function H_S. Let $f_{\mathrm{emp}}^M(\omega)$ be the empirical mean of a variable f of the subsystem, i.e.,

$$f_{\mathrm{emp}}^M(\omega) = \frac{1}{M} \sum_{i=1}^{M} f(\omega_{S_i}).$$

---

[8] We know that the stationarity requirement need not yield a unique measure. This should not be viewed as a failure of the typicality idea *per se*. In more technical terms, a typicality measure is really a representative of an equivalence class of measures, which are absolutely continuous with respect to each other. For stationary measures which are not absolutely continuous with respect to each other (like the microcanonical and canonical measures), we need to apply further insights, for example, that the universe is a "closed" system which does not exchange energy with some "outside" system, so that the total energy of the universe is fixed. In quantum mechanics, by the way, we shall have uniqueness of the typicality measure.

For example, the variable could be the kinetic energy of system i, viz.,

$$f(\omega_{S_i}) = \sum_{k=1}^{N} \frac{p_{S_i,k}^2}{2m_k} .$$

Alternatively, it could be the characteristic function of a set, viz., $f(x) = \mathbb{1}_A(x)$, in which case we consider the empirical distribution proper (see Definition 3.1 for an example). The justification of the statistical hypothesis, e.g., for the microcanonical ensemble, where we fix $H_S = E$, will then be formulated in terms of the law of large numbers, whence we return to Theorem 3.1:

Theorem 3.3 (Justification of the Statistical Hypothesis) *For all $\epsilon > 0$ and $\delta > 0$, there exists a number M such that, for all $n \geq M$,*

$$\mathbb{P}_{H_U}\left(\left\{\omega : \left|f_{\text{emp}}^n(\omega) - \int_{\Omega_E^S} f(\omega_S)\, d\mathbb{P}_E(\omega_S)\right| > \epsilon\right\}\right) < \delta .$$

This is the precise mathematical formulation of the statistical hypothesis stated above.

Remark 3.5 (On Explanations in Terms of Typicality) In Sect. 3.1, we already discussed the particularity of typicality assertions and we wish to elaborate a little on that. It is understandable to doubt at first whether a result like Theorem 3.3 qualifies as a proof of anything. In our usual understanding, a proof carves in stone the necessity of an assertion, it doesn't just say that something is typical, without even specifying where the borderline for typicality is fixed. But here, that just isn't on the cards. No law of nature prevents the coin from *always* landing on heads or forbids all the gas molecules in the lecture hall from moving simultaneously to the right half of the hall.

Those things are not impossible, but they are *atypical*. We might perhaps come up with the idea of showing that the statistical hypothesis holds true for the exact initial conditions of our universe, but clearly, we do not have access to those initial conditions. And even if we could specify at least one initial condition of the universe for which the statistical hypothesis were true, why should that be the one we live in? Why would that yield an *explanation* for our experiences?

The explanatory value arises only if we can show that the statistical regularities expressed by the statistical hypothesis do not just hold for some special initial conditions, but for the *overwhelming majority* of initial conditions, i.e., if we can show that it is typical. This reasoning may take some getting used to, but it is well worthwhile, since almost all macroscopic law-like behaviour is typical in this very sense.

Remark 3.6 (A More Relevant Formulation) Although it may be confusing at this point, we should point out that the formulation of the statement of the above theorem is still not quite on target. This formulation tacitly assumes that typical universes allow for subsystems of the kind we are discussing, in particular for ensembles of similar subsystems on which measurements of relative frequencies can be performed. But that may be an unwarranted assumption. In fact, the Boltzmann analysis suggests that typical universes are utterly boring, like a dilute homogeneous gas. In this case, universes containing subsystems of the kind we are interested in—like subsystems containing coins and people throwing them—would form a set of very small typicality measure. Therefore we should revise the theorem by considering the conditional measure, restricted to the set of universes containing subsystems and ensembles of the kind we are interested in. Only the conditional typicality measure can yield a relevant assertion. In fact, in the next chapter, in Theorem 4.1, we discuss the analogous theorem in Bohmian mechanics, but giving the relevant conditional form.

It is worth remarking that this type of theorem is not discussed in physics textbooks. Recall, however, our short remark on the ideal gas and the Maxwellian velocity distribution after the statement of the statistical hypothesis. The result described there can be formulated in connection with Theorem 3.3, where the precise result would involve a thermodynamic limit of infinitely many particles and hence needs to be taken with a grain of salt.[9] On the other hand, it is clear that it will in general be extremely difficult to design a proof for a universe which is much more complex than the ideal gas.

However, this is not the main reason why the theorem does not appear in such books. The main reason is that it does not seem to fit our experiences. In various situations, the statistical hypothesis, i.e., that we should find an equilibrium distribution, is indeed fulfilled, as in coin-tossing experiments, or the air filling a lecture hall, but we nevertheless experience plenty of everyday violations of equilibrium behaviour. We boil water for coffee, we cool water to make ice, we fill flasks with gas under high pressure, and in fact we ourselves in our human form constitute an assembly of molecules which seems altogether atypical. In fact, almost everything around us seems atypical. So atypical that we could easily have the impression that our whole universe is atypical. This is related to the notion of *entropy* and to the justification of the thermodynamic arrow.[10]

Quantum mechanics is special in the sense that Theorem 3.3 can be shown quite generally in a suitably adjusted manner, where the theoretical prediction for empirical frequencies of particle positions is given by the Born distribution considered earlier. This miraculous discovery will be discussed next.

---

[9]Compare the instructive computation in D. Dürr and S. Teufel, *Bohmian Mechanics. The Physics and Mathematics of Quantum Theory*. Springer, 2009.

[10]See, for instance, D. Lazarovici and P. Reichert, Arrow(s) of Time without a Past Hypothesis, in: V. Allori (ed.), *Statistical Mechanics and Scientific Explanation: Determinism, Indeterminism and Laws of Nature*, World Scientific, 2020 and S. Carroll: *From Eternity to Here*, Dutton, New York, 2010.

Bohmian Mechanics

4

> But in 1952 I saw the impossible done. It was in papers by David Bohm. Bohm showed explicitly how parameters could indeed be introduced, into non-relativistic wave mechanics, with the help of which the indeterministic description could be transformed into a deterministic one. More importantly, in my opinion, the subjectivity of the orthodox version, the necessary reference to the "observer," could be eliminated. Moreover, the essential idea was one that had been advanced already by de Broglie in 1927, in his "pilot wave" picture. But why then had Born not told me of this "pilot wave"? If only to point out what was wrong with it? Why did von Neumann not consider it? More extraordinarily, why did people go on producing "impossibility" proofs, after 1952, and as recently as 1978? When even Pauli, Rosenfeld, and Heisenberg could produce no more devastating criticism of Bohm's version than to brand it as "metaphysical" and "ideological"? Why is the pilot wave picture ignored in textbooks? Should it not be taught, not as the only way, but as an antidote to the prevailing complacency? To show us that vagueness, subjectivity, and indeterminism are not forced on us by experimental facts, but by deliberate theoretical choice?
>
> John S. Bell, *On the impossible pilot wave*[1]

As mentioned in Chap. 2, different quantum theories without observers have been suggested and we shall present one now which is particularly well worked out: Bohmian mechanics—a quantum theory of particles in motion.

To be clear from the very beginning, the empirical predictions of Bohmian mechanics are consistent with Born's statistical hypothesis, which is the basis for the empirical statements in standard quantum mechanics. So, wherever standard quantum mechanics makes unambiguous predictions, those agree with Bohmian predictions. In standard quantum mechanics, Born's statistical hypothesis is laid down as one of the axioms in the abstract quantum formalism, while the empirical import of Bohmian mechanics comes from a Boltzmannian typicality analysis which allows us to justify Born's statistical hypothesis. This can be seen as an example *par*

---

[1] Foundations of Physics **12**, 989 (1982). Reprinted in Bell, J. S. *Speakable and Unspeakable in Quantum Mechanics*, Cambridge University Press, Cambridge, 2004.

© Springer Nature Switzerland AG 2020

D. Dürr, D. Lazarovici, *Understanding Quantum Mechanics*,

https://doi.org/10.1007/978-3-030-40068-2_4

excellence of Boltzmann's idea of the statistical analysis of a deterministic theory, which we spelt out in the last chapter. Events will appear as if they are random, but in truth they are not.

Bohmian mechanics is not simply the "good old" quantum mechanics with point particles included in an *ad hoc* manner, or with particle trajectories included as supplementary variables. It is rather a proposal for a fundamental microscopic theory from which the quantum formalism can be derived as an effective statistical description, rather analogous to the way thermodynamics in the sense of statistical mechanics is derived from the classical mechanics of point particles. In the old Copenhagen interpretation, Bohr insisted that macroscopic objects such as the pointers of measurement devices need definite positions to record the outcome of measurements. The most straightforward way to achieve this consistently is to describe them as being made out of microscopic objects—particles—that always have definite positions. In Bohmian mechanics, there is thus no "shifty split" between a classical world of localized objects and a non-understandable quantum world. The Bohmian theory goes all the way: particles exist and move—also in the quantum world. Not a big deal, as it turns out.

The basic idea of this theory was already presented by Louis de Broglie at the famous 1927 Solvay conference, where some of the most distinguished physicists had gathered for a discussion about the contemporary understanding of nature. But his idea received a cool rejection.

De Broglie suggested that the wave function, which was at that time connected to matter in some "unspecified" way, could be thought of as a pilot wave for point particles which choreographed the motion of the particles. Only a few physicists, like for example Hendrik Lorentz (1853–1928), showed much sympathy for de Broglie's attempt to develop a theory of particle trajectories. In fact, in the general discussion session during that conference, Lorentz said:

> For me, an electron is a corpuscle that, at a given instant, is present at a definite point in space, and if I had the idea that at a following moment the corpuscle is present somewhere else, I must think of its trajectory, which is a line in space. [...] I imagine that, in the new theory, one still has electrons. It is of course possible that in the new theory, once it is well developed, one will have to suppose that the electrons undergo transformations. I happily concede that the electron may dissolve into a cloud. But then I would try to discover on which occasion this transformation occurs. [...] I am ready to accept other theories, on condition that one is able to re-express them in terms of clear and distinct images.[2]

This reflects Lorentz' wish to acquire a clear picture, just like de Broglie, but it also questions Bohr's school of thought regarding the wave–particle duality which was crystallising during that period. Bohr insisted that a quantum object could change its appearance like a chameleon—sometimes it would look like a wave and sometimes it would look like particle. But Lorentz was asking: What is the physical

---

[2]Quoted from G. Bacciagaluppi and A. Valentini, *Quantum Theory at the Crossroads: Reconsidering the 1927 Solvay Conference*, Cambridge University Press, Cambridge, 2009, p. 433.

process behind this change? In de Broglie's ansatz, on the contrary, there was always both a particle *and* a wave.

Neither Albert Einstein (1879–1955) nor Erwin Schrödinger showed any sympathy for de Broglie's project, which is quite remarkable because both opposed the central role of the observer in the Copenhagen interpretation of quantum mechanics. We shall explore the reason for Einstein's stance in Chap. 10. The situation changed from around 1950 onwards, with the work of David Bohm and John Bell. David Bohm, being ignorant of de Broglie's early attempts, came to the same idea quite independently, and almost completely developed the theory.

A crisp motto encapsulating the essence of Bohmian mechanics is this: *When we say "particle", we actually mean "particle"*. In quantum mechanics, there is endless talk about particles—for example electrons which are sent through a slit and subsequently strike a detector screen, as is clearly visible when a black spot appears on the photo screen. But as soon as we inquire about the use of the word "particle", we are told insistently that this is just a manner of speaking and should not be taken seriously. Perhaps the reader has learned that in quantum mechanics a particle cannot have a sharp position as long as it is not measured. Or that the uncertainty relation renders the existence of particle trajectories impossible. Or that an electron is sometimes a particle and sometimes a wave. All these statements are unjustified in the sense that they do not follow from the phenomena, as the mere existence of Bohmian mechanics clearly demonstrates.

If we acknowledge that quantum mechanics ought to be about physical entities in the world and if we take the notion of *point particle* seriously, it is easy to find a law of motion for the particles which is not only able to reproduce the phenomena of quantum physics, but which explains them and makes them understandable. That such a theory is not plagued by the measurement problem has already been understood in Chap. 2. According to Bohmian mechanics, at any given instant of time, every physical system has a well defined configuration of matter, given by the positions of the particles in three-dimensional space. After a measurement the pointer of a piece of measurement apparatus—consisting of Bohmian particles—points either to the right or to the left, while the wave function of the apparatus represents a superposition of both possibilities. All that remains for us to do is to understand the law of motion of the particles and see how the statistical predictions of quantum mechanics can be derived from that.

Bohmian mechanics is, like Hamiltonian mechanics, a theory of point particles in motion, but in contrast to Hamiltonian mechanics, it is not a Newtonian theory. In Bohmian mechanics the law governing the motion of N particles is not represented by a vector field on the $6N$-dimensional phase space but by a vector field on the $3N$-dimensional configuration space of the N particles, viz.,

$$\mathbb{R}^{3N} = \left\{ q \mid q = (\mathbf{q}_1, \ldots, \mathbf{q}_N), \ \mathbf{q}_k \in \mathbb{R}^3 \right\},$$

where the actual configuration is denoted by

$$Q := (\mathbf{Q}_1, \ldots, \mathbf{Q}_N), \quad \mathbf{Q}_k \in \mathbb{R}^3 .$$

In Hamiltonian mechanics, it is the Hamilton function on phase space which generates the vector field, while in Bohmian mechanics, it is Schrödinger's wave function ψ, which is defined not on phase space but on configuration space. In classical physics, we rarely consider cases where the Hamilton function is time-dependent, although those situations would be precisely the ones that would be important for technical applications like increasing the pressure in a gas by means of a piston. In quantum mechanics, we often consider wave functions which are time-dependent. Hence it seems reasonable to adjoin the wave function ψ to the state description of a Bohmian system. The state of a Bohmian system is thus given by a pair (ψ, Q) where ψ is the wave function and Q the particle configuration of the system.

Bohmian mechanics is thus defined by the evolution equation for the wave function and by the evolution equation of the particle configuration. The wave function

$$\psi : \mathbb{R}^{3N} \times \mathbb{R} \to \mathbb{C}, \quad \psi(q, t),$$

is taken to be a solution of the Schrödinger equation

$$i\hbar \frac{\partial}{\partial t} \psi(q, t) = - \sum_{k=1}^{N} \frac{\hbar^2}{2m} \Delta_k \psi(q, t) + V(q) \psi(q, t). \tag{4.1}$$

The role of the wave function ψ is to generate a vector field v^{ψ} on $\mathbb{R}^{3N}$. To see this, we write

$$\psi(q, t) = R(q, t) e^{\frac{i}{\hbar} S(q, t)}, \tag{4.2}$$

where R and S are real-valued functions. Taking the $3N$-dimensional gradient ∇, the vector field is

$$v^{\psi}(q, t) := \frac{1}{m} \nabla S(q, t). \tag{4.3}$$

The configuration trajectories are then integral curves along the vector field, which means that, for the evolution of the particles with positions $(\mathbf{Q}_1, \ldots, \mathbf{Q}_N) = Q \in \mathbb{R}^{3N}$, we have

$$\frac{d}{dt} Q(t) = v^{\psi}(Q(t), t). \tag{4.4}$$

Equivalently, we can express (4.3) in the form

$$
\begin{aligned}
v^\psi(q,t) &= \frac{\hbar}{m} \mathrm{Im} \nabla \ln\left(\psi(q,t)\right) \\
&= \frac{\hbar}{m} \mathrm{Im} \frac{\nabla \psi(q,t)}{\psi(q,t)} = \frac{\hbar}{m} \mathrm{Im} \frac{\psi^*(q,t) \nabla \psi(q,t)}{\psi^*(q,t) \psi(q,t)},
\end{aligned} \tag{4.5}
$$

where ψ^* is the complex conjugate of ψ. For the position of the k th particle, we get

$$
\frac{\mathrm{d}}{\mathrm{d}t} \mathbf{Q}_k(t) = \frac{\hbar}{m} \mathrm{Im} \frac{\nabla_k \psi(q,t)}{\psi(q,t)} \bigg|_{q=(\mathbf{Q}_1(t),...,\mathbf{Q}_N(t))}, \tag{4.6}
$$

where $\nabla_k = \partial/\partial \mathbf{q}_k$. Bohmian mechanics is thus defined by two equations: the Schrödinger equation for the wave function and the guiding equation for the evolution of the positions of the particles, in which the wave function defines the velocity vector field. Written compactly in configuration space notation, we have

$$
\mathrm{i}\hbar \frac{\partial}{\partial t} \psi(q,t) = -\frac{\hbar^2}{2m} \Delta \psi(q,t) + V(q)\psi(q,t) \quad \text{(Schrödinger equation)},
$$

$$
\frac{\mathrm{d}}{\mathrm{d}t} Q(t) = \frac{\hbar}{m} \mathrm{Im} \frac{\nabla \psi(q,t)}{\psi(q,t)} \bigg|_{q=Q(t)} \quad \text{(guiding equation)}.
$$

Here is another way to arrive at (or express) the velocity in the guiding law. The numerator of the rightmost term in (4.5) is

$$
\frac{\hbar}{m} \mathrm{Im}\left(\psi^*(q,t) \nabla_k \psi(q,t)\right) = \frac{1}{2\mathrm{i}}\left[\psi^*(q,t)\nabla_k\psi(q,t) - \psi(q,t)\nabla_k\psi^*(q,t)\right] = j^\psi(q,t),
$$

the quantum flux (1.5)! Hence, (4.5) reads

$$
v^\psi(q,t) = \frac{j^\psi(q,t)}{|\psi(q,t)|^2}, \tag{4.8}
$$

and we see that the Bohmian trajectories turn out to be the flux lines of the quantum flux. We shall return to this in Sect. 4.2, which discusses typicality.

Remark 4.1 ("Particles with Spin") If the wave function is a spinor wave function [see (1.26)] of the form

$$
\psi = \begin{pmatrix} \psi_1 \\ \psi_2 \end{pmatrix},
$$

we now read ψ^* as $\psi^+ = (\psi_1^*, \psi_2^*)$, where ψ_k^* is the complex conjugated spinor component. Hence,

$$\psi^*\psi := \psi^+\psi = \psi_1^*\psi_1 + \psi_2^*\psi_2$$

can be interpreted as a scalar product in spinor space. Expressions like this appear in (4.5) and can be interpreted in this way for spinor wave functions. We then immediately obtain the extension of the guiding law for particles guided by a spinor wave function. These are the ones that are usually described as "particles with spin". But in view of Remark 1.5, this yields only part of the extension! Recall that spin-1/2 particles like electrons are guided by spinor wave functions which obey the Pauli equation (1.28). The correct extension is therefore given by the replacement of the quantum flux in (4.8) by the Pauli flux $\mathbf{J}_{\text{Pauli}}^{\psi}$ [see (1.31)].

Remark 4.2 (Parameters in the Theory) At the outset, the quantities $\hbar$ and m are just dimensional quantities whose meaning only becomes clear *a posteriori* when we analyse the theory. In the definition of the velocity field and in the Schrödinger equation, we only need a parameter of dimension $[\text{length}]^2[\text{time}]^{-1}$, which takes care of the naturally existing dimensions of space and time. The fact that the parameter can then be usefully expressed as $\hbar/m$, with m as the Newtonian mass, is seen by looking at the way Newtonian mechanics is embedded in Bohmian mechanics. In this way the dimensional constant $\hbar$ acquires its meaning as a mediator between the new Bohmian mechanics and the old Newtonian mechanics which comes out as a limiting case of the former.[3]

Remark 4.3 (The Strange Transformation Property of the Wave Function) Noting that the wave function is complex-valued, the reader may wonder why complex numbers appear at such a fundamental level in quantum mechanics. Before the advent of quantum mechanics, complex numbers were abstract mathematical objects deprived of any physical relevance. But now they are needed to secure time reversal invariance. Time reversal $t \mapsto -t$ goes hand in hand with complex conjugation $\psi \mapsto \psi^*$: if the time changes sign in (4.6), a minus sign appears on the left, and it also does so on the right if ψ is replaced by ψ^*.

Should we be worried that ψ behaves in such a strange manner? No, because if we understand the role the wave function plays in the theory, then the transformation property becomes natural. In Bohmian mechanics, the role of the wave function is to determine the trajectories of the particles, and these must not change under time reversal. As a geometric curve, the trajectory remains the same, and it is only the way the curve is traversed that changes. This dictates how the wave function must behave under time reversal. An analogous situation exists in classical

---

[3]For more details see, e.g., D. Dürr and S. Teufel, *Bohmian Mechanics. The Physics and Mathematics of Quantum Theory*. Springer, 2009.

electromagnetism: the magnetic field **B** changes under time reversal to $-\mathbf{B}$, for the very same reason!

Remark 4.4 (Quantum Potential and Classical Limit) In general, the wave function which appears here as generator of the Bohmian velocity vector field obeys the time-dependent Schrödinger equation (1.2). The wave function thus generally depends on time. In contrast, we are not used to thinking of the Hamiltonian function in classical physics as the solution of a differential equation. But even in Hamilton's day, attempts were made to define mechanics on configuration space, in what has become known as Hamilton–Jacobi theory. In this approach, the so-called action function S, defined on configuration space, generates a vector field in exactly the same way as (4.3):

$$v = \frac{1}{m}\nabla S .$$

The action function S satisfies a partial differential equation which—and one should not be too surprised about this—is closely related to the Schrödinger equation. We elaborate on this because it leads straightforwardly to the embedding of classical mechanics in Bohmian mechanics as a limiting case. Using the polar form (4.2) in the Schrödinger equation (4.1) yields

$$i\frac{\partial R}{\partial t} - \frac{1}{\hbar}R\frac{\partial S}{\partial t} = -\frac{\hbar}{2m}\left[\Delta R + 2\frac{1}{\hbar}i\nabla R\cdot\nabla S - R\left(\frac{1}{\hbar}\nabla S\right)^2 + iR\frac{1}{\hbar}\Delta S\right] + \frac{R}{\hbar}V ,$$

where we omit the arguments for ease of notation. Sorting into real and imaginary parts, we obtain for the imaginary part

$$\frac{\partial R}{\partial t} = -\frac{\hbar}{2m}\left(2\frac{1}{\hbar}\nabla R\cdot\nabla S + R\frac{1}{\hbar}\Delta S\right) ,$$

or again

$$\frac{\partial R^2}{\partial t} = -\frac{1}{m}\nabla\cdot\left(R^2\nabla S\right) \overset{(4.3)}{=} -\nabla\cdot\left(v^\psi R^2\right) . \tag{4.9}$$

Since $R^2 = |\psi|^2$ this is again the quantum flux equation (1.4). Meanwhile, the real part reads

$$\frac{\partial S}{\partial t} - \frac{\hbar^2}{2m}\frac{\Delta R}{R} + \frac{1}{2m}(\nabla S)^2 + V = 0 . \tag{4.10}$$

The latter would be precisely the Hamilton–Jacobi equation for the action S if it weren't for the extra term

$$-\frac{\hbar^2}{2m}\frac{\Delta R}{R},$$

which Bohm called the quantum potential. If we think of the classical Hamilton–Jacobi equation, i.e., when the quantum potential is zero, as synonymous with classical mechanics, we can say that we get classical mechanics in the "classical limit", i.e., in situations where the quantum potential is negligible.

4.1   From the Universe to Subsystems

The quantum theories discussed in this book are non-relativistic theories and therefore describe our universe only incompletely and approximately. Nevertheless, when we wish to understand and analyse theories, we must take them as serious candidates for a fundamental description of nature. In this sense Bohmian mechanics is a serious theory, and as with any such theory, its defining law does not contain a limited region of validity in terms of, e.g., particle numbers or system size. The defining equations (4.1) and (4.6) are valid for arbitrarily big systems and hence also for the universe as a whole. Therefore, on the fundamental level, there exists in Bohmian mechanics—as in any other precise quantum theory—only *one* wave function, namely the *wave function of the universe*, which, in the case of Bohmian mechanics, guides *all* particles.

But the universe is too big for practical physics. If we couldn't describe an atom in our laboratory without taking into account the particles in the Andromeda nebula (or the particles of a nearby experimenter for that matter), the theory would indeed be useless. We must therefore ask how subsystems of a large system can be described "autonomously". The answer in quantum mechanics is different from the one given in Hamiltonian mechanics. This is due to the entanglement of the wave function on configuration space, which complicates the description of subsystems. In Hamiltonian mechanics, we can argue that interactions become small over large distances or that the force which acts on a subsystem from a more or less homogeneous distribution of distant matter is approximately zero. We may also find an effective Hamiltonian description of a subsystem using an "external" potential. Such arguments are not sufficient to achieve autonomous descriptions of subsystems in Bohmian mechanics, or in other quantum theories for that matter.

In view of (4.6), we conclude that, if the wave function of the total system (subsystem plus environment) is a product $\Psi(x, y) = \varphi(x)\Phi(y)$, and if this product structure remains intact for some time, then the system with wave function $\varphi(x)$ and the environment with wave function $\Phi(y)$ will evolve independently of each other; and hence the particles of the system with wave function $\varphi(x)$ and the particles in the environment with wave function $\phi(y)$ will move independently of each other. Alternatively, we may think of (4.2) and (4.3) and recall that the phases of the wave functions add if they form a product. But the assumption of a product form of the wave function is not generally justifiable. Interactions, no matter how small, will produce entanglement, and this will not decay over large distances. How then is it possible to achieve an autonomous description of subsystems?

In Bohmian mechanics, there is a quite natural answer because there are particles which define the subsystem! Suppose that the large system consists of N particles $(\mathbf{Q}_1, \ldots, \mathbf{Q}_N)$ and consider a subsystem of N_1 particles with positions $X = (\mathbf{Q}_1, \ldots, \mathbf{Q}_{N_1})$, so that the environment consists of the rest of the particles with positions $Y = (\mathbf{Q}_{N_1+1}, \ldots, \mathbf{Q}_N)$. The coordinates q in $\mathbb{R}^{3N}$ will therefore naturally split into

$$q = (x, y), \quad x \in \mathbb{R}^{3N_1}, \quad y \in \mathbb{R}^{3(N-N_1)}.$$

The question is now whether the x-system obeys its own Bohmian law. Will it have a wave function on its own? In principle, it is the large system which is guided by

$$\Psi(q, t) = \Psi(x, y, t). \tag{4.11}$$

On the fundamental level where the large system is the whole universe, that wave function is called the *universal wave function*. From that we obtain a function on the configuration space $\mathbb{R}^{3N_1}$ of the subsystem directly by setting

$$\varphi^Y(x, t) := \Psi(x, Y(t), t), \tag{4.12}$$

that is to say, by plugging in the actual coordinates $Y(t)$ of the particles making up the environment. The function (4.12) is called the *conditional wave function*, a term which will become clearer later on.

The conditional wave function is indeed the wave function of the x-system because, in view of (4.6), the reader may check that

$$\dot{X}(t) = v_x^\Psi(X(t), Y(t)) = \frac{\hbar}{m} \, \mathrm{Im} \frac{\nabla_x \Psi(x, Y(t), t)}{\Psi(x, Y(t), t)} \bigg|_{x=X(t)} = \frac{\hbar}{m} \, \mathrm{Im} \frac{\nabla_x \varphi^Y(x, t)}{\varphi^Y(x, t)} \bigg|_{x=X(t)}.$$

If the conditional wave function is required to be normalized, we take

$$\|\Psi(\cdot, Y)\| = \left(\int |\Psi(x, Y)|^2 \mathrm{d}^{3N_1} x \right)^{1/2} = 1.$$

The wave function of a subsystem is thus the conditional wave function. This conditional wave function behaves just like the wave function in standard quantum mechanics is supposed to. If the subsystem is sufficiently isolated, the conditional wave function obeys its own Schrödinger equation, while in a measurement experiment it "collapses" automatically, without interference from the "observer". Let us now see why this is.

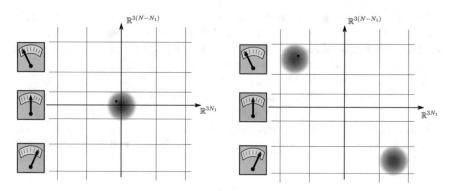

Fig. 4.1 Splitting of the wave function in configuration space in a "measurement process". The dot signifies a possible configuration of the Bohmian particles

Recalling our discussion of the measurement process in Chap. 2,

$$\varphi\psi_0 = (c_1\varphi_1 + c_2\varphi_2)\psi_0 \xrightarrow{\text{Schrödinger evolution}} c_1\varphi_1\psi_1 + c_2\varphi_2\psi_2, \qquad (4.13)$$

where $\varphi(x) = c_1\varphi_1(x) + c_2\varphi_2(x)$ is the wave function of the measured system, i.e., the x-system is in a superposition of two normalized wave functions which would lead in the measurement to results 1 and 2, respectively, and $\psi(y)$ is the wave function of the measurement apparatus, i.e., the y-system constitutes the "environment". ψ_0 is the ready state and ψ_i, $i = 1, 2$, are the wave functions corresponding to the two possible measurement results, i.e., pointer left corresponds to 1 and pointer right to 2. ψ_1 and ψ_2 represent macroscopically distinct states so they have macroscopically disjoint supports in configuration space (see Fig. 4.1).[4]

Suppose now that the Bohmian configuration Y of the measurement apparatus after completion of the measurement is in the support of ψ_1 (pointer left). By definition, the conditional wave function of the x-system is then

$$\varphi^Y(x) := c_1\varphi_1(x)\psi_1(Y) + c_2\varphi_2(x)\psi_2(Y) \approx c_1\varphi_1(x)\psi_1(Y), \qquad (4.14)$$

---

[4]We must take the notion of macroscopic disjointness with a grain of salt, since it will only be true in an approximate sense (although to an excellent approximation), e.g., in the sense of L^2, which means that wave functions are close if their $|\Psi|^2$-measures are close. In view of Born's law (see Sect. 4.2), this sense of closeness is the empirically relevant one. In short,

$$\Psi \approx \tilde{\Psi} \iff P^\Psi \approx P^{\tilde{\Psi}}.$$

since $\psi_2(Y) \equiv 0$, at least to a good approximation and for long times. When normalized, this is

$$\varphi^Y(x) = \varphi_1(x),\qquad(4.15)$$

where the equality is meant projectively, i.e., it holds up to a phase factor.

So the conditional wave function of a subsystem, which leads to an effective description of the subsystem, can collapse in the sense described here, but the universal wave function, or the wave function of the larger isolated system of which the subsystem is a part, does not collapse! It is important to keep this in mind, as it makes all the difference in the collapse models discussed in Chap. 5. In Bohmian mechanics, macroscopic interference is in principle possible, while in collapse models it is not.

If we plug the total wave function of the right-hand side of (4.13) into the guiding equation (4.5) and assume that $\psi_2(Y) \equiv 0$, we see that the velocity vector field of the x-system is determined solely by φ_1. The "empty" part of the total wave function $\propto \varphi_2(x)\psi_2(y)$ does not guide the system, and since both parts decohere, we may forget that part for all practical purposes. The remaining dependence on the environmental coordinates Y (via ψ_1) cancels out. Note that the argument does not require the total wave function to be a superposition of product wave functions as in (4.14). The following much weaker but still special condition suffices:

$$\Psi(x, y) = \varphi(x)\Phi(y) + \Psi^\perp(x, y),\qquad(4.16)$$

where Φ and $\Psi^\perp$ have macroscopically disjoint y-supports[5] and

$$Y \in \mathrm{supp}\,\Phi.\qquad(4.17)$$

In this case we still call $\varphi(x)$ the *effective wave function* of the subsystem. If the interaction term $V(x, y)\varphi(x)\Phi(y)$ can be neglected in the Schrödinger equation, at least for some time [use (4.1) with (4.11) and (4.16)], then the x-system and the environment are dynamically decoupled. The effective wave function subsequently obeys its own Schrödinger equation and the x-system behaves as an "isolated Bohmian system".

Once again, to get these ideas clear, the conditional wave function is precisely defined by (4.12) at all times, but it generally depends on the configuration of the environment. In fact, the conditional wave function generally evolves in a very complicated stochastic kind of way. In particular circumstances, however, the conditional wave function can become an effective wave function which allows an autonomous description of a Bohmian subsystem. Such special circumstances

---

[5] Although macroscopic disjointness is clearly a vague notion, it is nevertheless sufficiently intuitive to be practical. To understand macroscopically disjoint y-supports, a configuration space picture might be helpful.

are usually encountered in laboratories where isolated systems are "prepared" or "measured", but also in the "classical limit" when certain macroscopic systems can be treated as independent—for all practical purposes.

And to bring the message home, we stress once again that the conditional wave function (or the effective one) is precisely the representative of the "collapsed" wave function that is usually used in standard textbook quantum mechanics, and in particular in the description of measurements.

Remark 4.5 (Reduced Density Matrix) In the statistical description of a subsystem in standard quantum mechanics, we often consider the so-called *reduced density matrix*. To do this, we write the right-hand side of the total wave function (4.13) as the density matrix

$$\rho_{\text{total}} = |c_1|^2 |\varphi_1\rangle |\psi_1\rangle \langle\varphi_1| \langle\psi_1| + |c_2|^2 |\varphi_2\rangle |\psi_2\rangle \langle\varphi_2| \langle\psi_2| \tag{4.18}$$
$$+ c_1 c_2^* |\varphi_1\rangle |\psi_1\rangle \langle\varphi_2| \langle\psi_2| + c_1^* c_2 |\varphi_2\rangle |\psi_2\rangle \langle\varphi_1| \langle\psi_1| .$$

We then take the partial trace over the environmental degrees of freedom, whence they disappear from further considerations. Supposing that ψ_1 and ψ_2 are orthogonal vectors in the Hilbert space, which is the case if they have disjoint supports, we obtain the reduced density matrix of the x-system in the form

$$\rho_x = |c_1|^2 |\varphi_1\rangle\langle\varphi_1| + |c_2|^2 |\varphi_2\rangle\langle\varphi_2| . \tag{4.19}$$

The reader is encouraged to do this as an exercise. The disappearance of the non-diagonal elements in (4.18) is one way of making decoherence manifest (see Sect. 2.5). The reduced density matrix yields the correct *statistical* description for an *ensemble* of subsystems, as will be justified shortly, but it does not generally describe the actual state of a single subsystem. This is clear because, if *we see* for example the pointer pointing left and we can say with certainty that a repetition of the measurement yields the same result, then *we know* what the wave function is! In such situations standard quantum mechanics must invoke the collapse postulate to collapse the wave function to either φ_1 or φ_2. This may sound a bit awkward, but that's because the old quantum theory lacks precisely those elements which determine the *actual* state of the system.

4.2 Typicality Analysis and Born's Statistical Interpretation

In this section we draw heavily on the insights we gained from Sect. 3.5. We repeat here the analysis of that section for the new physical theory Bohmian mechanics. The reader is thus advised to revisit that section for the basic ideas behind Boltzmann's typicality analysis. We begin by asking: What is the meaning of (4.9), which we derived from Schrödinger's equation? This equation is in fact the continuity equation [see (3.17)] for the Bohmian flow T_t^ψ, as we shall now explain. T_t^ψ is generated by the solutions of (4.4)

$$T_t^\psi : \mathscr{Q} \longrightarrow \mathscr{Q}, \quad \mathscr{Q} = \mathbb{R}^{3N}. \tag{4.20}$$

The flow transports initial values $Q(0) = (\mathbf{Q}_1(0), \ldots, \mathbf{Q}_N(0))$ to values

$$Q(t) = \big(\mathbf{Q}_1(t; Q(0)), \ldots, \mathbf{Q}_N(t; Q(0))\big)$$

along Bohmian trajectories.

In many applications of classical mechanics, the trajectories are computed for initial values of interest, and it is tempting to do the same thing now for Bohmian trajectories. For example, in classical physics, by fixing the initial position and velocity, it is easy to compute the trajectory of a harmonic pendulum. It may be surprising to learn that our feeling that we can determine initial conditions with arbitrary precision is due to the global non-equilibrium of our universe, which we discussed in our investigation of chance in physics in Chap. 3. One new feature introduced in Bohmian mechanics is that the positions of Bohmian particles cannot be controlled with arbitrary precision, and this as a matter of principle: the particle positions are in *quantum equilibrium* with the wave function. This means that, given the wave function, we need only focus on *typical* initial values when using the theory to explain phenomena. Our aim is now to clarify these claims.

Recalling (4.8), we have

$$j^\psi(q, t) = v^\psi(q, t)|\psi(q, t)|^2, \tag{4.21}$$

so the quantum flux equation (1.4) can be written in the form

$$\frac{\partial |\psi|^2}{\partial t} = -\nabla \cdot v^\psi |\psi|^2. \tag{4.22}$$

This then shows that the quantum flux equation is the continuity equation for the Bohmian flow, where in the present setting the special density $|\psi|^2$ is transported along the velocity vector field (4.3). To be sure, the continuity equation for the Bohmian flow for an arbitrary density ρ is

$$\frac{\partial \rho}{\partial t} = -\nabla \cdot v^\psi \rho.$$

Equation (4.22) is the Bohmian counterpart of *Liouville's Theorem* 3.2 in Hamiltonian mechanics. We recognize $\rho = |\psi|^2$ to be a special density, namely one that does not change its form in ψ as time goes on. This is a natural generalisation of *stationarity*, called *equivariance* (see below). As in (3.20), for example, as a function of H, the density $e^{-\beta H}$ does not change with time and the same holds here. But the expression $|\psi|^2$ is so much simpler as a function of ψ!

The situation is actually more gratifying than in classical mechanics because this stationarity, or more precisely, this equivariance, determines $\rho = |\psi|^2$ uniquely as a stationary density. *Every* other density $\rho = f(\psi)$ with positive f would change in some way with the Bohmian flow, but never in such a way that $\rho(t) = f(\psi(t))$ holds, except for $f(x) = x^2$. In a nutshell, $\rho = |\psi|^2$ is unique.[6]

Remark 4.6 (Revisiting History) Equations (4.2)–(4.6) constitute a mathematical representation of de Broglie's idea back in 1927, viz., there are particles and they move along trajectories determined by the wave function. These trajectories are such that the theory seems—at least superficially—to agree with Born's statistical interpretation of the wave function. Why then was it so harshly rejected?

Here is a first criticism which at that time seemed very convincing, but which completely ignored Boltzmann's insights: $|\psi|^2$ is a probability density and probability is subjective degree of belief, i.e., ψ is merely an expression of our ignorance or degree of belief. It makes no sense to say that such a subjective quantity could actually move particles! Recalling Boltzmann's statistical analysis, however, this critique is well off target.

But there was another more striking criticism. ψ is a function on the configuration space of let's say N particles. The Schrödinger equation does not evolve a general ψ into one which has product structure, which would allow each particle to have its own guiding wave function, even when the particles move far away from each other. But this means that in general all particles will be guided by a common wave function—an *entangled* wave function. In view of this, Bohmian mechanics (or de Broglie's suggestion) is very far from classical physics, and in particular from relativistic physics. There is a kind of holism or non-separability taking place in physics, where all particles are interconnected by being guided by a common wave function. Einstein found this to be unacceptable and referred to such a ψ as a "Geisterfeld" (ghost field).

Einstein's criticism of this kind of holism, in which the wave function acts in an "eerie" manner, needs to be taken seriously. John Stewart Bell showed just how serious: Einstein put his finger on the key innovation of quantum mechanics. The wave function does act in precisely this holistic way. It acts this way because Nature is nonlocal! We shall discuss this in detail in Chap. 10. Put another way, Bohmian mechanics is nonlocal, just as Nature requires it to be.

Let us now proceed with the typicality analysis in Bohmian mechanics. We consider a very large Bohmian system (a universe) as the dynamical system, taking N as the dimension of configuration space and Ψ as the wave function generating the flow (4.20). We thus begin with the triple

$$(\mathscr{Q}, T_t^{\Psi}, \mathbb{P}^{\Psi}),$$

---

[6]See S. Goldstein und W. Struyve, On the uniqueness of quantum equilibrium in Bohmian mechanics. Journal of Statistical Physics **128** (5), 1197–1209 (2007).

where $\mathbb{P}^\Psi$ is the measure of typicality, assumed to have the property of *equivariance* which generalizes the notion of stationarity for time-dependent vector fields. We express the property of equivariance once more in terms of the time translated measure to bring out the full analogy with Boltzmann's typicality analysis:

$$\mathbb{P}_t^\Psi(A) := \mathbb{P}^\Psi \circ (T_t^\Psi)^{-1}(A) = \mathbb{P}^\Psi\left((T_t^\Psi)^{-1}(A)\right) = \mathbb{P}^{\Psi_t}(A), \qquad (4.23)$$

or more generally, for arbitrary functions f,

$$\int f(Q(t))\mathrm{d}\mathbb{P}^\Psi = \int f(Q)\mathrm{d}\mathbb{P}^{\Psi_t}.$$

Diagramatically, this can be represented by

$$\begin{array}{ccc}
\Psi & \longrightarrow & \mathbb{P}^\Psi \\
\Big\downarrow{\scriptstyle U(t)} & & \Big\downarrow{\scriptstyle \circ(T_t^\Psi)^{-1}} \\
\Psi_t & \longrightarrow & \mathbb{P}^{\Psi_t}
\end{array}$$

Here, the downward arrow on the left-hand side indicates the unitary Schrödinger time evolution $U(t)$ of the wave function, while the downward arrow on the right-hand side represents $\mathbb{P}^\Psi \mapsto \mathbb{P}_t^\Psi$, i.e., the time evolution of the measures given by transport along the Bohmian trajectories.

So which measures $\mathbb{P}^\Psi$ satisfy that? As we said above: the only solution is the typicality measure

$$\mathbb{P}^\Psi(A) = \int_A |\Psi(q)|^2 \, \mathrm{d}^{3N} q, \qquad (4.24)$$

with the normalisation

$$\int |\Psi(q)|^2 \, \mathrm{d}^{3N} q = 1, \qquad (4.25)$$

if normalisation is required. The typicality measure determines (in Boltzmann's view) the empirical distributions in the Bohmian universe. How do we get the empirical distributions? For that we need an ensemble of identical (or very similar) subsystems. We already know how to describe subsystems in Bohmian mechanics, namely, using the conditional wave function. We split the configuration space as we already did previously by dividing the coordinates into two sets:

$$q = (x, y), \quad x \in \mathbb{R}^{3N_1}, \quad y \in \mathbb{R}^{3(N-N_1)},$$

where the x-degrees of freedom (dimension $3N_1 =: m$) describe the subsystem. The conditional wave function is then

$$\varphi^Y(x, t) := \Psi\left(x, Y(t), t\right),$$

where $Y(t)$ is the actual configuration of the environment, i.e., of the rest of the universe. In view of (4.24), this leads us directly to the conditional typicality measure:

$$\mathbb{P}^\Psi\left(\{Q = (X, Y) : X \in d^m x\} \mid Y\right) =: \mathbb{P}^\Psi\left(X \in d^m x \mid Y\right)$$

$$= |\Psi(x, Y)|^2 d^m x$$

$$= |\varphi^Y(x)|^2 d^m x. \qquad (4.26)$$

Thus, by conditioning the typicality measure on the actual configuration of the environment, we obtain once again a measure of the form $\mathbb{P}^\Psi$, where we need only replace the density by the one given by the conditional wave function.

Remark 4.7 This is all intuitively clear, but mathematics requires rigour, and alas, we conditioned with $y = Y$, which is a set of measure zero. For mathematical rigour, we would need to bring in derivatives of measures. We only mention this in passing. For the present discussion, it would not be worth going any further into this matter.

Equation (4.26) does indeed incorporate the empirical import of Bohmian mechanics, but a few more steps are needed. The specification of the actual coordinate Y of the complete environment of the x-system is much too detailed, and this would in fact be practically useless. Likewise, the formula itself may not look particularly useful in this form. But the right-hand side of (4.26) depends solely on the conditional wave function and that allows a coarse-graining of the environment: we can coarse grain so long as we can guarantee that the given conditional wave function is not affected. This means that we can coarse grain so much that we condition finally only on the event that the conditional wave function φ^Y equals φ. So, let

$$\mathscr{Y}^\varphi := \left\{Q = (X, Y) : \varphi^Y = \varphi\right\}$$

be the set of Q for which the conditional wave function equals φ. Equation (4.26) then implies the simpler, directly applicable formula

$$\mathbb{P}^\Psi\left(\{Q = (X, Y) : X \in d^m x\} \mid \mathscr{Y}^\varphi\right) = |\varphi|^2 d^m x. \qquad (4.27)$$

Equation (4.27) follows from (4.26) by a simple property of conditional measures: for $B = \bigcup B_i$ a pairwise disjoint family with $\mathbb{P}(A|B_i) = a$ for all B_i, we have

$$\mathbb{P}(B)a = \sum_i \mathbb{P}(A|B_i)\mathbb{P}(B_i) = \sum_i \mathbb{P}(A \cap B_i) = \mathbb{P}(A \cap B),$$

and thus $\mathbb{P}(A|B) = \mathbb{P}(A \cap B)/\mathbb{P}(B) = a$.

We now consider empirical distributions and the law of large numbers. Here, we shall obtain a proof of Born's statistical interpretation, and we shall also show what it actually means! We consider an ensemble $X = (X_1, \ldots, X_M)$ of similar subsystems $X_1, \ldots, X_M$, which evolve independently of each other. Do such subsystems exist? The answer is that they obviously do exist, at least in our universe! This is exactly the way experiments are performed! Hence the conditional wave function of the ensemble must be a product of the wave functions φ_i of each subsystem. Otherwise the subsystems would not be independent of each other.[7] So, we have

$$\varphi^Y(x_1, x_2, \ldots, x_M) = \prod_{i=1}^{M} \varphi_i(x_i),$$

and therefore the measure of typicality conditioned on the event of the ensemble being described by the above turns out to be[8]

$$\mathbb{P}^\psi\left(\{Q = (X, Y) : X_1(Q) \in d^m x_1, \ldots, X_M(Q) \in d^m x_M\} \mid \mathcal{Y}^\varphi\right) = \prod_{i=1}^{M} |\varphi(x_i)|^2 d^m x_i .$$

$$(4.28)$$

Note that we are now looking at a product measure, a measure under which the law of large numbers becomes a triviality, as can be seen from the proof of Theorem 3.1 in the more general setting.

Suppose for concreteness that we are interested in the typical value of the empirical distribution for the position of a particle having conditional wave function φ. Hence, the ensemble $X = (X_1, \ldots, X_M)$ consists of M similar particles, each with wave function φ. The empirical distribution of the particle positions

---

[7] Actually, it suffices to require that every member of the ensemble have effective wave function φ_i, from which the product structure of the ensemble wave function then follows. See, for example, D. Dürr and S. Teufel, Bohmian Mechanics. The Physics and Mathematics of Quantum Theory. Springer, 2009.

[8] Note that the subsystem coordinates $X_1, \ldots, X_M$ are coarse-graining functions of the universal coordinates Q, where the universal configuration space is the fundamental Ω space in the sense of probability theory.

$X_1, \ldots, X_M$ in the ensemble is then (see footnote 8)

$$\rho_{\mathrm{emp}}^M(A, Q) := \frac{1}{M} \sum_{k}^{M} \mathbb{1}_A\big(X_k(Q)\big) . \tag{4.29}$$

More generally, instead of $\mathbb{1}_A$, we may consider a general function f, as we did in Theorem 3.3. Then, using (4.28) and the law of large numbers, we immediately obtain the justification of what is accordingly called the quantum equilibrium hypothesis:

Theorem 4.1 (Quantum Equilibrium Distribution)

$$\mathbb{P}^\Psi\left(\left\{Q : \left|\frac{1}{M} \sum_{i=1}^{M} f(X_i) - \int f(x)|\varphi(x)|^2 \mathrm{d}x\right| < \varepsilon\right\} \,\middle|\, \mathscr{Y}^\varphi\right) = 1 - \delta(\varepsilon, f, M) ,$$

$$\tag{4.30}$$

where $\delta(\varepsilon, f, M)$ gets arbitrarily small with increasing M.

Here f can be any coarse-graining function of the particle coordinates. In particular, for $f = \mathbb{1}_A$ with $A \subset \mathbb{R}^3$, the theorem predicts the typical value for the empirical distribution (4.29), i.e., the relative frequency of positions, viz.,

$$\rho_{\mathrm{emp}}^M(A, Q) \approx \int_A |\varphi(x)|^2 \mathrm{d}x .$$

In words, the quantum equilibrium hypothesis can be stated as follows:

> If a system has wave function φ, then the coordinates of the particles of the system are $|\varphi|^2$ distributed , $\tag{4.31}$

where "distributed" refers to the empirical distribution. This is in fact what is meant by Born's statistical interpretation of the wave function.

It is a speciality of quantum mechanics that the typicality measure and the typical empirical distributions always have the form of a "wave-function-squared measure", i.e., where the density has the form $|\Phi|^2$, the former with respect to the universal wave function, the latter with respect to the conditional wave function. This is mathematically nice and helpful, but didactically somewhat unfortunate. Indeed, without having digested Boltzmann's argumentation, we may have the impression that we have presented a "garbage-in garbage-out argument", in the sense that we stick in Born's statistical distribution and we get out Born's statistical distribution, i.e., $|\Phi|^2$ in, $|\Phi|^2$ out! How can that have any explanatory or predictive power?

We thus repeat: $\mathbb{P}^\Psi$, i.e., $\Phi = \Psi$, is the typicality measure whose role is to define *typical* initial conditions for the Bohmian universe. That measure is distinguished from all other measures by stationarity (or equivariance). For typical initial conditions, i.e., typical universes, the empirical distribution ρ_{emp} is given by Born's statistical distribution $|\varphi|^2$, i.e., $\Phi = \varphi$, and applies to subsystems. This implies the well known empirical predictions of quantum mechanics. Note also that Born's statistical rule in the form of, let's say, (4.31), which gives ρ_{emp} in terms of the conditional or effective wave function (in terms of the "collapsed" wave function as it is called in orthodox quantum mechanics) is neither postulated nor used as "input"—it is proven.

The Bohmian Theorem 4.1 is the prototype of a Boltzmannian typicality statement, but note that we always need to quantify the function δ, otherwise the theorem has no practical use. In view of Remark 3.6, we can say that we have now formulated Theorem 4.1 in the proper and relevant way, viz., the *conditional* typicality measure $\mathbb{P}^{\Psi}(\{\ldots\}|\mathscr{Y}^{\varphi})$ of the configurations for which Born's law does not hold is small. We repeat the point made in the remark. If it was only the $\mathbb{P}^{\Psi}$-measure for the deviation that was small, that would not mean much because the configurations to which our relevant environment Y belongs may already have small measure. It is important that the predictions are made for all the relevant environments in which there exist experiments of the kind we are interested in with appropriately prepared wave functions. This is achieved by conditioning.

The quantum equilibrium hypothesis (4.31) is a precise version of Born's statistical interpretation of the wave function (1.2), which is taken as an axiom in the quantum mechanics of the old school. By its justification in Bohmian mechanics, the role played by probability in quantum mechanics is now clear and beyond doubt: it is the same role as in classical mechanics.

In Bohmian mechanics, Theorem 4.1 is a consequence of typicality which has become known as *quantum equilibrium.*[9] The primary role of the wave function in Bohmian mechanics is not to compute probabilities but to guide particles. And because typicality should be determined by the physical law itself, it is the wave function which determines the typicality measure. Experimentally, no violation of the quantum equilibrium hypothesis has ever been found. We can safely assume that quantum equilibrium holds without exceptions, i.e., that our universe is a typical Bohmian universe.

Remark 4.8 (Absolute Uncertainty) This also means that we have a *principled* ignorance of the position of a particle. If its (conditional) wave function is φ, we cannot know or control the position of the particle better than is allowed by the $|\varphi|^2$-distribution. We can also understand this as follows. All the information we have about particle positions must be stored or recorded in the configuration Y of the rest of the universe, either written on paper or on the hard disk of a computer or in our brain. This information has already been taken care of when we construct the conditional measure (4.26). If a system with conditional wave function φ has been prepared, then we cannot know more about its particle configuration than is given

---

[9]D. Dürr, S. Goldstein, and N. Zanghì, Quantum equilibrium and the origin of absolute uncertainty. In: *Quantum Physics Without Quantum Philosophy*, Springer, 2013.

by the $|\varphi|^2$-distribution. This principled ignorance—called *absolute uncertainty*—should not be confused with the notion of intrinsic randomness. At each instant of time, the law of motion governs the particle positions deterministically. Absolute uncertainty is merely a consequence of typicality.

4.3   Heisenberg's Uncertainty

Each and every student of physics has heard about Heisenberg's uncertainty relation. It says that we cannot measure with arbitrary precision the velocity and the position of a particle at the same time. That by itself would be rather harmless, but it is also often said that this has a drastic consequence, namely that it makes no sense to even talk about the position and velocity of a particle, i.e., it makes no sense to even talk about a particle trajectory. But Bohmian mechanics is about exactly that: particle trajectories. How can that situation live alongside Heisenberg's uncertainty relation? Since this question may be justified from a historical perspective, we shall elaborate on it here.

The uncertainty relation is a direct consequence of Born's statistical interpretation of the wave function (1.2) and the spreading of the wave function (see Sect. 1.3), and of course of the Bohmian trajectories which describe the facts of the matter. We shall repeat that at the end of this section, but first we shall explain why it is so. To do this, we connect the long time asymptotic of the wave function with the Bohmian trajectories.

First of all, we recall what is involved in a momentum (or velocity) measurement in quantum mechanics (see Sect. 1.3). The simplest way is this. We measure the particle's position at time $t = 0$, let's say, and then let the wave function evolve freely, recording where the particle has got to at some later time t. We then divide the distance travelled by the time t. That ratio times the mass is the measured momentum value.

Here we do the computation in Bohmian mechanics. Let $\mathbf{X}_0$ be the particle's position at $t = 0$, and $\mathbf{X}(t, \mathbf{X}_0)$ its position at time t, where $\big(\mathbf{X}(t, \mathbf{X}_0)\big)_{t \geq 0}$ denotes the trajectory starting at $\mathbf{X} = \mathbf{X}_0$. Then for large t,

$$\frac{\mathbf{X}(t, \mathbf{X}_0) - \mathbf{X}_0}{t} \approx \mathbf{V}_\infty$$

yields the asymptotic velocity $\mathbf{V}_\infty$, and $m\mathbf{V}_\infty$ would be the classical momentum. Let us now compute how $\mathbf{V}_\infty$ is distributed.

If the wave function at $t = 0$ is ψ_0 (located around let's say 0), then $\mathbf{X}(0)$ is $|\psi_0|^2$ distributed. Next we need the distribution of

$$\frac{1}{t}\big[\mathbf{X}(t, \mathbf{X}_0) - \mathbf{X}_0\big] \approx \frac{1}{t}\mathbf{X}(t, \mathbf{X}_0),$$

where the approximation holds for large t. Hence we must calculate

$$\mathbb{P}^\psi \left(\frac{\mathbf{X}(t, \mathbf{X}_0)}{t} \in A \right)$$

for large t. However, that expression has already been computed in Sect. 1.3 using (1.11), and is given by (1.12). Hence, by defining the product of the mass and the asymptotic Bohmian velocity as the momentum, we get

$$\lim_{t \to \infty} \mathbb{P}^\psi \left(m \frac{\mathbf{X}(t, \mathbf{X}_0)}{t} \in A \right) = \int \mathbb{1}_A (\hbar \mathbf{k}) \left| \hat{\psi}_0(\mathbf{k}) \right|^2 d^3 k , \qquad (4.32)$$

where the right-hand side was already interpreted as the quantum mechanical momentum distribution in Sect. 1.3.

But let us get back to the uncertainty relation. If the particle has the wave function ψ, then the variance in its position is

$$\text{Var}(\mathbf{X}) = \mathbb{E}^\psi (\mathbf{X}^2) - \left[\mathbb{E}^\psi (\mathbf{X}) \right]^2 .$$

As we have just seen, the momentum inherits from ψ the variance

$$\text{Var}(\mathbf{P}) = \mathbb{E}^\psi (\mathbf{P}^2) - \left[\mathbb{E}^\psi (\mathbf{P}) \right]^2$$
$$= \int (\hbar \mathbf{k})^2 \left| \hat{\psi}(\mathbf{k}) \right|^2 d^3 k - \left(\int \hbar \mathbf{k} \left| \hat{\psi}(\mathbf{k}) \right|^2 d^3 k \right)^2 .$$

We consider a Gaussian wave function, centred around zero for simplicity. Then $|\psi|^2$ is a Gaussian distribution with, let's say, width σ. As a consequence, $\hat{\psi}$ will be Gaussian and so also will be $|\hat{\psi}|^2$, but with width $1/4\sigma$ (a good exercise for the reader). Not forgetting the factor $\hbar$, we thus get the uncertainty relation for Gaussians in the form

$$\left[\text{Var}(\mathbf{X}) \right]^{1/2} \left[\text{Var}(\mathbf{P}) \right]^{1/2} = \frac{\hbar}{2} . \qquad (4.33)$$

For general wave functions, a bit more analysis is needed to obtain the general form of Heisenberg's uncertainty relation in which the equals sign in (4.33) is replaced by $\geq$. At risk of repeating ourselves, we stress once again what is really going on here. The fact is that, the better one localises the particle at time $t = 0$, the more sharply the wave function is peaked (Born's rule) and the more (and higher) Fourier modes are needed to represent that wave function. The Fourier modes are plane waves (it is better to think of packets) which travel with speeds given by their wave numbers, so the more sharply peaked the initial wave function, the more widely the Fourier modes will separate. That's all there is to it.

The moral is that the uncertainty relation is a rather simple consequence of Bohmian mechanics and quantum equilibrium. Two insights should be taken home. First, the observable momentum does not generally correspond to the instantaneous velocity of the Bohmian particle as given by (4.4). Secondly, there is a difference between *empirical quantities* and *ontology*, that is, a difference between what can be measured and what exists according to the physical theory (see also Sect. 7.3).

4.4 Identical Particles and Topology

Identical particles, sometimes also referred to as indistinguishable particles, often lead to confusion. It is often said that identical particles in quantum physics are categorically different from identical classical particles in that they cannot even exist as discrete entities. Indistinguishability suggests such a view because a particle has a position (at least if one is reasonable enough to admit that), so particles are distinguished by their different positions. In Bohmian mechanics, particles exist, so does this contradict quantum mechanics? In fact, it does not. We only need to be clear about what is meant by indistinguishable particles. Furthermore, the quantum mechanical description of identical particles is a consequence of Bohmian mechanics!

So what is meant by identical particles? Really nothing else than that the comfortable (alas, we humans like comfort) enumeration of particles like $\mathbf{Q}_i$, $i = 1, \ldots, N$, plays no physical role. In other words, the physical law governing the motion of identical particles is ignorant of any labelling. Numbering is alright if it goes along with different properties of the particles, like the different masses in Newtonian physics which enter the law of motion as parameters. We nevertheless often use labelling because it allows a simpler way of speaking and sometimes also simpler mathematics. But if this is a no-go for identical particles, how can we actually handle them?

A very simple and often used argument for handling identical particles in quantum mechanics is the following. When particles are identical, Born's statistical interpretation of the wave function must respect that. This means that, if we exchange particle labels, the probability for finding one particle here and one there must not change, i.e., for two particles, this is expressed by requiring

$$|\psi(\mathbf{q}_1, \mathbf{q}_2)|^2 = |\psi(\mathbf{q}_2, \mathbf{q}_1)|^2 . \tag{4.34}$$

Hence the wave function ψ can only change by a phase factor under particle exchange, since that drops out when taking the absolute square. However, this argument does not apply to spinor wave functions, where a little more work is required to come to the same conclusion. But here we shall stick to complex-valued wave functions for simplicity.

Can more be said about the phase factor? Is it arbitrary? No, it is in fact either $+1$ or -1, whence

$$\psi(\mathbf{q}_1, \mathbf{q}_2) = \pm\psi(\mathbf{q}_2, \mathbf{q}_1).\qquad(4.35)$$

This means that only symmetric or antisymmetric wave functions can occur. Particles with symmetric wave functions are known as bosons, while those with antisymmetric wave functions are called fermions. We shall show (4.35) later and discuss first why the argument leading to (4.34) begs the following question: Isn't it strange to handle identical particles by first labelling them with numbers to eventually add that the numbering was nonsense to begin with, and hence kick them out again?

Why not work with identical particles at the outset? This insight came relatively late (in the 1970s) because it referred to the configuration space of identical particles, and in quantum physics with its denial of the existence of particles, this was not on the cards to begin with.[10] Of course, in Bohmian mechanics, configuration space is there in a quite natural way, but what exactly it looks like for identical particles needs to be spelt out. Because the proper handling of identical particles is a rather deep issue, we shall take the trouble to present it in some detail.

In order to render the numbering unimportant at the outset, we need to replace our beloved configuration space $\mathbb{R}^{3N}$, which represents the set of N-tuples of particle positions, by something rather different. Tuples are by definition ordered, i.e., their entries are distinguished by their order of appearance, and hence automatically numbered. To replace this kind of object, consider the following. We have N particles in $\mathbb{R}^3$, so we have N points in $\mathbb{R}^3$, which means that they form a subset of $\mathbb{R}^3$ with N elements. Now, sets are not ordered. A set is by definition a collection of *different* objects which can be distinguished by indices, but the sequence in which we count the elements plays no role, i.e., any other numbering will yield the same set. Therefore the configuration space for N identical particles can be expressed as

$$\mathscr{Q} = \left\{q \subset \mathbb{R}^3 : |q| = N, \text{ where } q = \{\mathbf{q}_1, \ldots, \mathbf{q}_N\}, \ \mathbf{q}_i \in \mathbb{R}^3\right\},$$

where $|q|$ is the cardinality of the set q. This set of subsets of a given finite cardinality—actually a manifold—looks fairly simple at first sight, but with some scrutiny, we find that the manifold is in fact rather complicated from a topological point of view. The first thing to note, however, is that a wave function defined on this

---

[10]Concerning its role at the beginning of quantum physics, de Broglie said in the Solvay conference *Électrons et Photons*, Paris 1928, p. 111: *Il semble un peu paradoxal de construire un espace de configuration avec des coordonnées de points qui n'existent pas.* (We leave the translation to the reader.) It should be noted that the configuration space of identical particles is an old concept, sitting right at the heart of Boltzmann's statistical mechanics. But in classical physics the topology of the configuration space plays little role, contrary to the situation in quantum physics, as we shall see. In fact, it is the entanglement of the wave function that changes the picture.

manifold is in a sense intrinsically symmetric. The argument of the wave function is a set, which does not change when we permute particle coordinates.

We can express the wave function $\psi(q)$ as a function of the tuples $\mathbf{q}_i$, $i = 1, \ldots, N$, of the elements of $q \in \mathcal{Q}$ by introducing them as coordinates of $\mathcal{Q}$. A change in the order of appearance in the tuple corresponds to a change of coordinates, where the configuration point q remains the same. Hence, $\psi(\mathbf{q}_1, \ldots, \mathbf{q}_N)$ is invariant under each permutation σ of the N indices:

$$\psi(\mathbf{q}_{\sigma(1)}, \ldots, \mathbf{q}_{\sigma(N)}) = \psi(\mathbf{q}_1, \ldots, \mathbf{q}_N).$$

Thus only symmetric wave functions are possible, i.e., the phase factor mentioned above is always $+1$.

So what about Pauli's exclusion principle for identical particles, which says that a many-particle wave function cannot contain a product of two or more identical one-particle wave functions. The symmetric wave functions violate this principle, but it is fulfilled by antisymmetric wave functions

$$\psi(\mathbf{q}_{\sigma(1)}, \ldots, \mathbf{q}_{\sigma(N)}) = \text{sign}(\sigma)\psi(\mathbf{q}_1, \ldots, \mathbf{q}_N),$$

where $\text{sign}(\sigma)$ is the sign of the permutation, i.e., -1 if σ is a product of an uneven number of transpositions and $+1$ if their number is even. How can we obtain those? By thinking about the topology!

What is the topological nature of $\mathcal{Q}$? This is best understood by looking at the following equivalent description of the configuration space of N identical particles:

$$\mathcal{Q} = (\mathbb{R}^{3N} \setminus \Delta^{3N})/S_N =: \mathbb{R}^{3N}_{\neq}/S_N.$$

In this expression, $\neq$ instructs us to remove the "diagonal"

$$\Delta^{3N} := \left\{ (\mathbf{q}_1, \ldots, \mathbf{q}_N) \in \mathbb{R}^{3N} \,|\, \mathbf{q}_i = \mathbf{q}_j \text{ for at least one } i \neq j \right\}, \qquad (4.36)$$

from $\mathbb{R}^{3N}$, yielding the set $\mathbb{R}^{3N}_{\neq} := (\mathbb{R}^{3N} \setminus \Delta^{3N})$. We then consider all N-tuples in $\mathbb{R}^{3N}_{\neq}$ which are permutations of each other as equivalent or identical:

$$(\mathbf{q}_{\sigma(1)}, \ldots, \mathbf{q}_{\sigma(N)}) \sim (\mathbf{q}_1, \ldots, \mathbf{q}_N).$$

This construction of equivalence classes gives a so-called factorisation by the permutation group S_N of N objects, which is denoted by $\mathbb{R}^{3N}_{\neq}/S_N$.

We met a similar situation in Sect. 1.8 when we considered the Lie group $SO(3)$ as a manifold by identifying points and their antipodes on the surface of a solid sphere. The moral is once again that, by identifications of this kind, closed paths may no longer be null-homotopic and manifolds may no longer be simply connected. Such manifolds are not so easy to handle mathematically. Indeed, they hide some interesting possibilities which can be revealed by "unfolding" the identifications, by going to the universal covering, as discussed briefly at the end of Sect. 1.8 for the example of the covering of $SO(3)$ by $SU(2)$.

A simple example is the circle Q constructed from the interval $[0, 1]$ by identifying 0 with 1, or from $\mathbb{R}$ by identifying all $n \in \mathbb{Z} \subset \mathbb{R}$ with zero. The latter gives $\mathbb{R}/\sim$, where $x \sim y$ iff $x - y \in \mathbb{Z}$. $\mathbb{R}$ is topologically as simple as one can get, but the circle is not. It contains closed paths which cannot be shrunken to a single point homotopically, i.e., in a continuous manner or, loosely speaking, without breaking the path. Indeed, any path going right around the circle at least once cannot be so transformed. The set of closed paths on the circle can be divided into equivalence classes which can be indexed by the number of windings (positive or negative). A closed path going around the circle, let's say, three times positively and once negatively is equivalent to the closed path which goes around twice positively.

We can define a "product" of closed paths by joining them together. For example, adjoining any path to the same path but in the opposite direction, we obtain the null path, the path which stays put. In this manner, we can consider the closed paths as elements of a group. The group is called the *fundamental group* Π of the manifold. For the circle, viewed as a manifold, this group is $\Pi \cong \mathbb{Z}$. If the fundamental group is not trivial, i.e., if it is not isomorphic to a group with only one element e such that $e \times e = e$, where $\times$ denotes the group operation, the manifold is said to be *multiply connected* rather than *simply connected*.

The identification may be undone, by unwinding the simply connected manifold $\hat{Q} = \mathbb{R}$ in a spiral above the circle Q, observing that successive elements of length 2π are mapped bijectively onto the circle (see Fig. 4.2). Such a construction is called a covering, and when the covering space is simply connected, as it is in our example, it is called the *universal covering*. This is mathematically much simpler to work with.

Analogously, $\mathcal{Q} = \mathbb{R}^{3N}_{\neq}/S_N$ is multiply connected, and in this case S_N is isomorphic to the fundamental group. To see why, take two particles in $\mathbb{R}^3$ and the point $(\mathbf{q}_1, \mathbf{q}_2)$ in configuration space (as depicted in Fig. 4.3). This has to be identified with $(\mathbf{q}_2, \mathbf{q}_1)$, i.e., those two points are one and the same point in $\mathbb{R}^6_{\neq}/S_2$.

Fig. 4.2 The circle Q is covered by $\hat{Q} = \mathbb{R}$. The fibre is the set of all points $q \cong \hat{r} \cong \hat{s} \cong \cdots \in \hat{Q}$ which are to be identified with the point $q \in Q$. The elements of the covering group $Cov(\hat{Q}, Q)$ (see text) map points to each other within one fibre. It is isomorphic to the fundamental group Π of Q, which is isomorphic to $\mathbb{Z}$ in the case of the circle

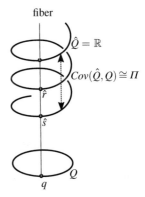

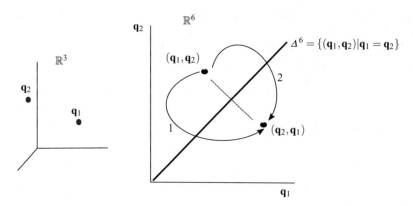

Fig. 4.3 *Left*: Two identical particles in $\mathbb{R}^3$. *Right*: Sketch of the configuration space of the two particles in $\mathbb{R}^6$, where the diagonal Δ^6 has been removed. The points above the diagonal are to be identified (using the notation $\sim$) with the mirror image points lying below the diagonal. Since $(\mathbf{q}_1, \mathbf{q}_2) \sim (\mathbf{q}_2, \mathbf{q}_1)$, both paths 1 and 2 are simply closed. Neither path 1 nor path 2 can be continuously deformed to the null path, because the removed diagonal constitutes a topological barrier (see text). In contrast, the doubly closed path $1 \circ 2$ is null-homotopic as explained in the text

Hence the path going from $(\mathbf{q}_1, \mathbf{q}_2)$ to $(\mathbf{q}_2, \mathbf{q}_1)$ in $\mathbb{R}^6$ without crossing the diagonal Δ^6 [see (4.36)] is a closed path in $\mathbb{R}^6_{\neq}/S_2$. Suppose we try to shrink this closed path homotopically to a point. This means continuously deforming the path without breaking it. We may imagine moving the "endpoint" $(\mathbf{q}_2, \mathbf{q}_1)$ towards $(\mathbf{q}_1, \mathbf{q}_2)$, but in $\mathbb{R}^6_{\neq}/S_2$ the two points are the same, so that does not work without breaking the path! We thus see that such a closed path is not null-homotopic.

But now adjoin a closed path from $(\mathbf{q}_2, \mathbf{q}_1)$ to $(\mathbf{q}_1, \mathbf{q}_2)$ to the closed path from $(\mathbf{q}_1, \mathbf{q}_2)$ to $(\mathbf{q}_2, \mathbf{q}_1)$. This yields a doubly closed path which in contrast can be continuously deformed, because there is now only one "endpoint". For example, we can now shift $(\mathbf{q}_2, \mathbf{q}_1)$ towards a point $(\mathbf{q}'_2, \mathbf{q}'_1)$ which lies closer to $(\mathbf{q}_1, \mathbf{q}_2)$, a procedure which would have torn the previous closed path apart, the added path keeps the whole path intact and the whole path is still closed, although $(\mathbf{q}_1, \mathbf{q}_2) \sim (\mathbf{q}'_2, \mathbf{q}'_1)$. But we still need to make sure that the diagonal Δ^6 presents no hindrance to shrinking the path to a null path.

To understand why it does not, the following analogy is helpful. How much of the configuration space is taken away by the removal of the diagonal? The diagonal is three-dimensional and the configuration space is six-dimensional, i.e., one still has three dimensions outside the diagonal. An analogous situation is encountered in $\mathbb{R}^3 \setminus \{0\}$, i.e., the three-dimensional space with the origin taken out. Consider in this case a closed path "around the origin", say a closed path in the (x, y)-plane, surrounding the origin, one can simply continuously lift that path in the z-direction, whereupon it can be shrunk to zero because the hole at the origin is no longer an obstacle. The same holds for our two-particle example, where we have enough extra dimensions into which the path can be "lifted" to avoid the obstacle constituted by the missing diagonal. In short, doubly closed paths are null-homotopic, while certain simply closed paths are not. The fundamental group thus has only two elements, the null path and the simply closed path we discussed. This corresponds to the permutation group with two elements, say $\{1, 2\}$, i.e., S_2.

To make sure that the picture is properly understood, it is a useful exercise to carry out the same reasoning for two particles in $\mathbb{R}^2$ and the corresponding configuration space $\mathbb{R}^4_{\neq}/S_2$. We now observe, again by the proper analogy, that the "diagonal" presents a barrier. However, it now leads to a much more complicated fundamental group than S_2. The new group is the so-called braid group, which enlarges the class of wave functions of fermions and bosons to what are called anyons. However, since we live in a three-dimensional space, the latter do not play such a fundamental role, and come into play only in approximations or effective descriptions, e.g., in solid state physics. But this remark is somewhat premature since we have not yet introduced the idea of bosonic and fermionic wave functions.

To get there, we start with the true configuration space of identical particles, viz., $\mathscr{Q} = \mathbb{R}^{3N}_{\neq}/S_N$, and as we did with the circle (Fig. 4.2), consider the simply-connected covering $\hat{\mathscr{Q}} = \mathbb{R}^{3N}_{\neq}$. As for the circle, we think of it as a spiral staircase. Moving from a point $\hat{q} = (\mathbf{q}_1, \ldots, \mathbf{q}_N)$ one flight up along a fibre, we arrive at the point $(\mathbf{q}_{\sigma(1)}, \ldots, \mathbf{q}_{\sigma(N)})$ with permuted coordinates. One fibre contains all points which are equivalent to $q \in \mathscr{Q}$, i.e., they are permuted coordinate representations that project to the same point in the "true" configuration space. We thus understand that we have $N!$ floors in this case, because there are $N!$ permutations. The floors on this stairway provide local coordinates for the basis manifold $\mathscr{Q}$.

There is now a second group, the so-called covering group $Cov(\hat{\mathscr{Q}}, \mathscr{Q})$. An element of that group maps $\hat{\mathscr{Q}}$ to $\hat{\mathscr{Q}}$, keeping the fibres invariant (see Fig. 4.2). In other words, the element maps one point on a fibre to another point on that fibre. Equivalently, for any two points $\hat{s}$ and $\hat{r}$ in the same fibre, which are thus projected along the fibre to the same point in $\mathscr{Q}$, there is always one element $\Sigma \in Cov(\hat{\mathscr{Q}}, \mathscr{Q})$, such that $\hat{s} = \Sigma \hat{r}$. Since in our case the fibre consists of permuted N-tuples it is clear that $Cov(\hat{\mathscr{Q}}, \mathscr{Q})$ is isomorphic to the permutation group S_N and hence to the fundamental group of $\mathscr{Q}$.

We now have the topological structure of the true configuration space $\mathscr{Q}$ in our hands and we can turn back to the physics. We would like to define the Bohmian velocity field on $\mathscr{Q}$. For this purpose, we consider wave functions $\hat{\psi}$ on $\hat{\mathscr{Q}}$, but noting that they must respect the fibre symmetry so that the vector field $\hat{v}^{\hat{\psi}}$ generated by $\hat{\psi}$ can also be viewed as a vector field on the basis manifold $\mathscr{Q}$. The technical way of putting this is that the vector field must be projectable to $\mathscr{Q}$. This means that the wave functions must satisfy a periodicity condition. It is intuitively clear what this condition must be if we recall the form of the vector field (4.5): for points $\hat{r}$ and $\hat{s}$ on a fibre with $\Sigma \hat{r} = \hat{s}$, where $\Sigma \in Cov(\hat{\mathscr{Q}}, \mathscr{Q})$, we must have

$$\hat{\psi}(\Sigma \hat{q}) = \gamma_\Sigma \hat{\psi}(\hat{q}) \,, \tag{4.37}$$

that is, the wave function can only change by a factor $\gamma_\Sigma \in \mathbb{C} \setminus \{0\}$. The constant factor by which $\hat{\psi}$ changes must be consistent with the group operation in the covering group, i.e.,

$$\hat{\psi}(\Sigma_2 \circ \Sigma_1 \hat{q}) = \gamma_{\Sigma_2} \gamma_{\Sigma_1} \hat{\psi}(\hat{q}) \,.$$

If we now also require that $|\gamma_\Sigma|^2 = 1$, i.e., $|\gamma_\Sigma| = 1$, we can project the equivariant evolution of the Bohmian trajectories in the covering space to the evolution in the true configuration space, so that the probability density

$$|\hat{\psi}(\Sigma \hat{q})|^2 = |\gamma_\Sigma|^2 |\hat{\psi}(\hat{q})|^2 = |\hat{\psi}(\hat{q})|^2$$

is projected to $|\psi(q)|^2$ defined on $\mathscr{Q}$.

In this way, we obtain a *unitary representation of the group*, which is also called a character representation. Translated into coordinates, this becomes

$$\hat{\psi}(\Sigma \hat{q}) = \hat{\psi}(\mathbf{q}_{\sigma(1)}, \ldots, \mathbf{q}_{\sigma(N)}) = \gamma_\Sigma \hat{\psi}(\mathbf{q}_1, \ldots, \mathbf{q}_N) \,, \qquad (4.38)$$

where we have expressed an element of the covering group by a permutation. We have now arrived at the point where the usual textbook presentations begin, e.g., starting with (4.34), so we can quicken the pace and reach a conclusion.

Consider (4.38) when σ is a transposition τ, exchanging exactly two indices. Since $\tau \circ \tau = \mathrm{id}$ and hence $\gamma_\tau^2 = 1$, it follows that $\gamma_\tau = \pm 1$. There are thus two character representations of the permutation group: one with $\gamma_\tau = 1$, the bosonic wave functions, and one with $\gamma_\tau = -1$, the fermionic wave functions.[11]

Remark 4.9 (Spinors and the Spin–Statistics Theorem) We close with some comments on spinor wave functions. Here, too, we have the boson–fermion alternative, but in order to show that, we need not only topology, but also dynamics, e.g., the Pauli equation with magnetic fields.[12]

We should also mention the Pauli principle which in modern form is known as the spin–statistics theorem. This states that particles with half integer spin are fermions, while those with integer spin are bosons. This theorem contains the boson–fermion alternative, but the coupling to half integer or integer spin has a dynamical origin. This can be seen by looking at the Dirac equation, the relativistic generalisation of the Pauli equation. It turns out that the "energy spectrum" of the Dirac equation, which governs the wave functions of electrons, is not bounded from below. This

---

[11]To be precise, we still need to prove that in a representation either $\gamma_\tau = 1$, $\forall \tau$, or $\gamma_\tau = -1$, $\forall \tau$. This is true because the permutation group can in fact be generated by elements of the form $\tau \circ \tau_0 \circ \tau$ for a fixed τ_0.

[12]For further details we refer to D. Dürr, S. Goldstein, J. Taylor, R. Tumulka, and N. Zanghì, Topological factors derived from Bohmian mechanics, Ann. H. Poincaré **7**, 791–807 (2006), reprinted in D. Dürr, S. Goldstein, and N. Zanghì, *Quantum Physics Without Quantum Philosophy*, Springer, 2013.

is annoying because electrons can in principle lose energy endlessly by radiating it away in the form of electromagnetic radiation. This scenario is known as the radiation catastrophe. If we now consider a universe with "infinitely many" fermions of negative energy, we can arrange them in such a way that no other electron can achieve negative energy, using the fact that two fermions, because of antisymmetry, cannot have the same wave function. We shall discuss this idea, due to Paul Dirac (1902–1984), in more detail in Chap. 11.

In the spin–statistics theorem, which is proven using quantum field theory, the infinitely many fermions do not occur explicitly. They are concealed in language like "vacuum", "particle", and "antiparticle". However, the assumptions made to prove the theorem always aim in some way or other to prevent the radiation catastrophe.

Concluding this chapter, Bohmian mechanics would be an almost trivial solution of the measurement problem, were it not for the indispensable configuration space which is not so easy to grasp in a pictorial manner.[13] All that Bohmian mechanics does is to take the word "particle" seriously. By doing so, all the mysteries of quantum orthodoxy not only evaporate, but we can also justify Born's rule, explain the role of operator observables (see Chap. 7 on the measurement process and observables[14]), and explain the fermion/boson alternative, and all this in the context of a theory that is manifestly nonlocal, as any theory of nature must be (see Chap. 10 on nonlocality).

A question often raised is why we cannot access the position of a Bohmian particle with arbitrary precision when its wave function is some arbitrary φ. To appreciate the answer we urge the reader to assimilate Remark 4.8 on absolute uncertainty.

---

[13]The German word often used here is "unanschaulich".

[14]For more on this, see D. Dürr and S. Teufel, Bohmian Mechanics. The Physics and Mathematics of Quantum Theory. Springer, 2009.

Collapse Theory

5

He [Schrödinger] would have liked, I think, that the theory is completely determined by the equations, which do not have to be talked away from time to time. He would have liked the complete absence of particles from the theory, and yet the emergence of 'particle tracks', and more generally of the 'particularity' of the world, on the macroscopic level. He might not have liked the GRW jumps, but he would have disliked them less than the old quantum jumps of his time. And he would not have been at all disturbed by their indeterminism.

John S. Bell, *Are there quantum jumps?*

The term "collapse theory" refers to an entire class of quantum theories that replace the Schrödinger equation by a non-linear time evolution for the wave function so that the superposition principle, which leads to the measurement problem, is no longer valid. For small systems, like single atoms, the violation of the superposition principle is barely noticeable. On the other hand, superpositions of macroscopic wave functions, such as for Schrödinger's cat, become practically impossible, or, more precisely, collapse so quickly that they can never be observed. Simply put, the wave function thereby collapses automatically with the right (that is, quantum mechanical) probabilities onto one of the localized wave packets associated with a well-defined macroscopic state. This process is also called *spontaneous localization*. It is helpful to recall our discussion of the measurement problem and Eqs. (2.4) to (2.6) in Sect. 2.4, where we already mentioned the possibility of such a theory solving the measurement problem.

What is crucial in all of these theories is that the process of wave function collapse is described by a precise mathematical law that holds always and for all systems. It is not introduced as a special power or property of the "observer" or "measurement process" as in orthodox quantum mechanics.

We say there is an "entire class" of such theories because, when it comes down to the details, there are many possibilities to modify the Schrödinger equation to include a collapse law. In the following, we will present the first and simplest type, which was introduced by Ghirardi, Rimini, and Weber (GRW theory) in the 1980s

© Springer Nature Switzerland AG 2020

D. Dürr, D. Lazarovici, *Understanding Quantum Mechanics*,

https://doi.org/10.1007/978-3-030-40068-2_5

and which John Bell also discussed. It was Bell who proposed the *flash ontology* for the GRW dynamics.[1]

The GRW theory describes the collapse dynamics as a discrete stochastic process (a so-called Poisson process). This means that the collapse occurs randomly, and the center of localization is (essentially) distributed according to the quantum mechanical probabilities, i.e., the $|\psi|^2$-density. We will make this more precise below. Nowadays, people usually study continuous spontaneous localization theories (CSL),[2] but their formulation requires mathematical tools (stochastic differential equations) that go beyond the scope of this book. There are even ideas to relate the collapse process to the influence of gravity, so that the collapse probability increases with the strength of the gravitational field, but these are somewhat speculative.[3]

Since collapse theories renounce the superposition principle, they make predictions that differ, in principle, from those of other quantum theories. These differences are very difficult to discern experimentally, but there is currently great interest in the kind of tests that might eventually confirm or falsify spontaneous collapse.[4] To this end, we can, for instance, do interferometer experiments with ever bigger molecules—essentially sending them through a sort of double-slit—since the violation of the superposition principle would become significant only for large systems. If collapse theories are correct, interference patterns should not occur above a certain object size, as spontaneous localization ensures that only one part of the superposition survives. (We will explain how it does that in a moment.) In this context, it is also easy to understand why experimental tests of spontaneous collapse are so challenging. Just recall our discussion of the influence of measurements (and the environment) in Sect. 1.4. The bigger the system, the harder it is to prevent decoherence due to interactions with the environment. This makes it very difficult to decide what is ultimately responsible for the disappearance of interference patterns: the theoretical collapse mechanism or environmental decoherence.

Spontaneous collapse could, however, reveal itself through certain side-effects. Collapse theories predict, for instance, that spontaneous localization leads to a random (Brownian) motion of the "particles", causing spontaneous heating and emission of radiation which astrophysicists can look out for. These effects are very weak and thus also difficult to detect, but there is hope that they will prove to be more accessible than direct violations of the superposition principle. In any case, the experimental data we have so far is insufficient to decide between spontaneous collapse and a linear Schrödinger evolution as presupposed by Bohmian mechanics or the Many Worlds theory. Of course, this may change in the foreseeable future.

---

[1]The term "flash" first appeared in a paper by Roderich Tumulka [A relativistic version of the Ghirardi–Rimini–Weber model. Journal of Statistical Physics **135** (4), 821–840 (2006)]. According to Tumulka, it was originally proposed by Nino Zanghì.

[2]See A. Bassi and G.C. Ghirardi, Dynamical reduction models. Physics Report **379**, 257 (2003).

[3]See, e.g., L. Diósi, Models for universal reduction of macroscopic quantum fluctuations. Physical Review A **40**, 1165 (1989), and R. Penrose, On the Gravitization of Quantum Mechanics 1: Quantum State Reduction. Foundations of Physics **44** (5), 557–575 (2014).

[4]See S.L. Adler and A. Bassi, Is Quantum Theory Exact?, Science **325** (5938), 275–276 (2009).

5.1 GRW Theory

Let's now take a closer look at the GRW theory and see how the collapse mechanism is put to work. First, a preliminary warning: it is common to talk about "particles" in the presentation of the theory, but this is just a way of speaking that cannot be taken too seriously. It merely refers to the fact that the wave function of a system is defined on $\mathbb{R}^{3N}$, $N \in \mathbb{N}$, which is isomorphic to the configuration space of N particles. "Particle k" thus refers first and foremost to the degree of freedom $\boldsymbol{q}_k$ in the wave function. In collapse theories, this choice of domain is *ad hoc*, although justified by its success in describing phenomena.

The GRW theory is now defined as follows:

1. As in standard quantum mechanics, the state of an N "particle" system is described by a wave function $\psi = \psi(\boldsymbol{q}_1, \ldots, \boldsymbol{q}_N, t)$. Between two collapse events, this wave function evolves according to the usual Schrödinger equation. The wave function changes instantaneously, however, whenever a collapse occurs.
2. For each particle $\boldsymbol{q}_k$, $k \in \{1, \ldots, N\}$, there is an independent probability for the occurrence of a collapse event. We denote such an event by $(T, \boldsymbol{X})_k$, where T is the random time and $\boldsymbol{X} \in \mathbb{R}^3$ the random location of the center of collapse. Following Bell, we call the collapse events *flashes*.

We now have to specify *how* the collapse acts, *when* it occurs, and *where* it occurs:

3. With the occurrence of a flash $(T, \boldsymbol{X})_k$, the wave function collapses in such a way that the parts far away from $\boldsymbol{X}$ are suppressed. More precisely, the collapse changes the wave function according to

$$\psi \longrightarrow \frac{\psi_{\boldsymbol{X}}^k}{\|\psi_{\boldsymbol{X}}^k\|}, \qquad \psi_{\boldsymbol{X}}^k := \mathrm{L}_{\boldsymbol{X}}^k \psi. \tag{5.1}$$

Here,

$$\mathrm{L}_{\boldsymbol{X}}^k := \left(\frac{1}{\pi a^2}\right)^{3/4} \exp\left[-\frac{(\boldsymbol{q}_k - \boldsymbol{X})^2}{2a^2}\right] \tag{5.2}$$

is the localization operator acting by multiplication by a Gaussian function centered around $\boldsymbol{X}$. The width a of the Gaussian appears as a new constant of nature (like c or $\hbar$) called the *localization width*. As far as we know today, it's order of magnitude is around $a = 10^{-7}$ m. Finally, the collapsed wave function is normalized according to

$$\|\psi_{\boldsymbol{X}}^k\|^2 = \int_{\mathbb{R}^{3N}} \mathrm{d}^3 q_1 \cdots \mathrm{d}^3 q_N \left|\mathrm{L}_{\boldsymbol{X}}^k \psi(\boldsymbol{q}_1, \ldots, \boldsymbol{q}_N, t)\right|^2. \tag{5.3}$$

The example below will illustrate why multiplication by a Gaussian amounts to a collapse or localization of the wave function.

4. The time between two collapse event is random and described by the "waiting time distribution"

$$\mathbb{P}(T \in dt) = \lambda e^{-\lambda t} dt \tag{5.4}$$

for each particle independently of all the others. That is to say, the probability for any given particle to trigger a collapse in the time interval Δt is

$$\int_0^{\Delta t} \lambda e^{-\lambda t} dt = 1 - e^{-\lambda \Delta t} .$$

The *collapse rate* λ is another new constant of nature, whose order of magnitude is assumed to be roughly $10^{-15}\,\mathrm{s}^{-1}$. For a single particle, the average time between two collapse events is thus $\approx 10^{15}$ sec, meaning that an isolated electron basically never flashes at all (the estimated age of the universe is roughly 10^{17} s). For a system with N particles, however, the collapse rate is $N/10^{15}\,\mathrm{s}^{-1}$, and when the scale of N is macroscopic, let's say $N \sim 10^{24}$, a number of flashes will occur within nanoseconds and destroy macroscopic superpositions before they can be perceived.

5. The probability of a flash (i.e., the center of collapse) of particle k occurring within the volume element d^3x is given by $\rho_k(x)\,d^3x$ with the density

$$\rho_k(\boldsymbol{x}) := \int_{\mathbb{R}^{3N}} d^3q_1 \cdots d^3q_N \left| L_{\boldsymbol{x}}^k \psi(\boldsymbol{q}_1, \ldots, \boldsymbol{q}_N, t) \right|^2 . \tag{5.5}$$

Note that the center $\mathbf{x}$ of the Gaussian $L_{\mathbf{x}}^k$ is the free variable. In total, the probability distribution of a $(T, \mathbf{X})_k$-flash is thus

$$\mathbb{P}\left(\{T \in (t, t+dt); \mathbf{X} \in d^3x\}\right) = \lambda e^{-\lambda t} \rho_k(\boldsymbol{x})\, d^3x\, dt .$$

5.2   Spontaneous Localization

We will now provide some examples to illustrate how the collapse acts and affects the wave function. For simplicity, we first consider a single particle in one dimension. Suppose that initially the wave function is widely spread, let's say

$$\psi(x) = \frac{1}{\sqrt[4]{\pi\sigma^2}} e^{-\frac{(x-x_0)^2}{2\sigma^2}} , \tag{5.6}$$

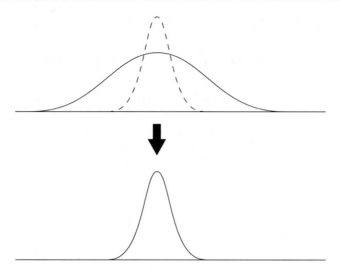

Fig. 5.1 Spontaneous localization of a wide wave packet

where $\sigma \gg a$, and that the collapse hits roughly at the center of the wave packet, i.e., $X \approx x_0$. Then, the wave function will change to

$$\psi_X = \frac{1}{N} \exp\left[-\frac{(x-x_0)^2}{2a^2}\right] \exp\left[-\frac{(x-x_0)^2}{2\sigma^2}\right]$$

$$\approx \frac{1}{N} \exp\left[-\frac{(x-x_0)^2}{2(a^2+\sigma^2)}\right] \approx \frac{1}{\sqrt[4]{\pi} a} \exp\left[-\frac{(x-x_0)^2}{2a^2}\right],$$

where N denotes the normalization factor. A spatially extended ψ thus becomes a collapsed wave function which is pretty well localized (with width a) around the center of collapse x_0 (see Fig. 5.1).

Now consider a superposition of two Gaussian wave packets centered around $\pm x_0$, $x_0 > 0$ with width σ:

$$\psi(x) = \frac{1}{N} \left\{\exp\left[-\frac{1}{2\sigma^2}(x+x_0)^2\right] + \exp\left[-\frac{1}{2\sigma^2}(x-x_0)^2\right]\right\}, \qquad (5.7)$$

where N is once again the normalization. We assume that $\sigma \ll a \ll x_0$, that is, the localization width a is much greater than the width of the Gaussians but smaller than the distance between their centers, i.e., the separation of the superposed wave packets. Suppose now that a collapse occurs at time $t = 0$. We want to compute the probability that the flash hits near x_0, let's say in the interval $(x_0 - 3\sigma, x_0 + 3\sigma)$.

According to (5.5) and (5.3), this probability is given by

$$\int_{(x_0-3\sigma,\,x_0+3\sigma)} \rho(x)\,dx = \int_{(x_0-3\sigma,\,x_0+3\sigma)} \int_{\mathbb{R}} |L_x\psi(y)|^2 dy\,dx\,. \qquad (5.8)$$

The y-integral on the right-hand side is a convolution of two Gaussian functions and thus easy to compute, and indeed we already did this in (1.9). The result is again a Gaussian whose mean and variance are the sum of those of the convoluted functions. We can make our lives even simpler, though. As our collapse operator L_x, respectively, L_x^2 is a very narrow Gaussian, convolution with L_x acts more or less like convolution with a delta function. We can thus approximate[5]:

$$\int L_x^2 |\psi(y)|^2 dy = \int \frac{e^{-(x-y)^2/a^2}}{\sqrt{\pi a^2}} |\psi(y)|^2 dy \approx \int \delta(x-y)|\psi(y)|^2 dy = |\psi(x)|^2\,,$$
$$(5.9)$$

whence the spatial distribution of the flashes is described by the Born rule. Inserted in (5.8), this yields

$$\int_{(x_0-3\sigma,\,x_0+3\sigma)} \rho(x)\,dx \approx \int_{(x_0-3\sigma,\,x_0+3\sigma)} |\psi(x)|^2 dx \approx 1/2\,,$$

since the left-hand wave packet (the one centered around $-x_0$) contributes almost nothing to the integral.

In conclusion, the flash has a chance of (roughly) $1/2$ to hit near x_0, the center of the right-hand wave peak, and analogously, a chance of (roughly) $1/2$ to hit near $-x_0$, the center of the left-hand peak. The flash may of course hit somewhere in-between, but the probability is almost zero.

So let's say the flash occurs at $X \approx x_0$. Then, the wave function changes as follows:

$$\psi(x) \longrightarrow \psi_X(x)$$

$$= \frac{1}{N_X} e^{-\frac{1}{2a^2}(x-X)^2} \left[e^{-\frac{1}{2\sigma^2}(x+x_0)^2} + e^{-\frac{1}{2\sigma^2}(x-x_0)^2} \right]$$

$$\underset{X\approx x_0}{\approx} \frac{1}{N_{x_0}} \left[e^{-\frac{(2x_0)^2}{2a^2}} e^{-\frac{1}{2\sigma^2}(x+x_0)^2} + e^{-\left(\frac{1}{2\sigma^2}+\frac{1}{2a^2}\right)(x-x_0)^2} \right]$$

$$\underset{\sigma\ll a}{\approx} \frac{1}{N_{x_0}} \left[e^{-\frac{(2x_0)^2}{2a^2}} e^{-\frac{1}{2\sigma^2}(x+x_0)^2} + e^{-\frac{1}{2\sigma^2}(x-x_0)^2} \right]\,,$$

---

[5]This approximation is exact in the limit $a \to 0$.

with the normalization factor $N_X := \|\psi_X^k\|$. We see that the part of the wave function centered around $-x_0$ is suppressed by the exponential factor $e^{-(2x_0)^2/2a^2}$ (note that $x_0 \gg a$), while the part centered around x_0 is amplified accordingly. And since the probability distribution for the localisation of the *next* flash is now determined by the collapsed wave function, it is very likely that it will further amplify the large branch and suppress the other (and so on).

Since the probabilities for the reduction of the wave packets correspond to a good approximation to those given by Born's rule, the statistical predictions of the GRW theory will be very close to those of standard quantum mechanics. To emphasize this point, we can read the above example as referring to a macroscopic wave function, describing, let's say, the state of a measurement device at the end of an experiment. It is then clear that, due to the very high collapse rate (many particles!), the wave function will collapse almost instantaneously into one of the possible pointer states with the right quantum mechanical probabilities. However, in contrast to the one-dimensional example, we now have to imagine the wave packets on the high-dimensional configuration space, as illustrated in Fig. 5.2.

From the previous calculation, we can also easily read off the case $a \gg x_0$, when the distance between the two wave packets is smaller than the localization width. In this case, the factor $e^{-(2x_0)^2/2a^2}$ is roughly 1, meaning that the collapse barely changes the wave function at all (see Fig. 5.3). The upshot is that the spontaneous collapse affects superpositions only on scales above the localization width a, while "small" superpositions remain intact.

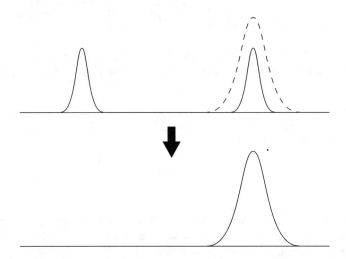

Fig. 5.2 Spontaneous collapse of a spatially separated superposition

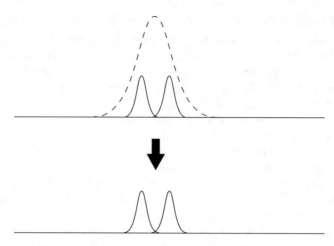

Fig. 5.3 "Small" superpositions are not destroyed by spontaneous collapse

For the next example, we consider two entangled particles in one dimension (generalization to N particles is then pretty straightforward):

$$\psi(x, y) = \frac{1}{N} \left\{ \exp\left[-\frac{(x + x_0)^2}{2\sigma^2}\right] \exp\left[-\frac{(y + y_0)^2}{2\sigma^2}\right] \right.$$
$$\left. + \exp\left[-\frac{(x - x_0)^2}{2\sigma^2}\right] \exp\left[-\frac{(y - y_0)^2}{2\sigma^2}\right] \right\}. \qquad (5.10)$$

Suppose the first collapse is triggered by particle 2, i.e.; in the y degrees of freedom, with center of collapse $Y \approx y_0$. An analogous computation to the one above then yields

$$\psi_Y^2 \approx \frac{1}{N} \left\{ \exp\left[-\frac{(2y_0)^2}{2a^2}\right] \exp\left[-\frac{(x + x_0)^2}{2\sigma^2}\right] \exp\left[-\frac{(y + y_0)^2}{2\sigma^2}\right] \right.$$
$$\left. + \exp\left[-\frac{(x - x_0)^2}{2\sigma^2}\right] \exp\left[-\frac{(y - y_0)^2}{2\sigma^2}\right] \right\}. \qquad (5.11)$$

Note that the exponential factor $e^{-(2y_0)^2/2a^2}$ comes from the flash of particle 2 but suppresses the left-hand wave packet of *both* entangled particles. On configuration space, we can visualize the collapse as shown in Fig. 5.4.

We can read this once again as a model for "large" systems, for instance, a microscopic system coupled to a macroscopic measurement device. Since the measurement device has many more degrees of freedom, it is very likely to trigger the collapse events, but these collapses will then also suppress superpositions of the microscopic system with which it is entangled at the end of the measurement

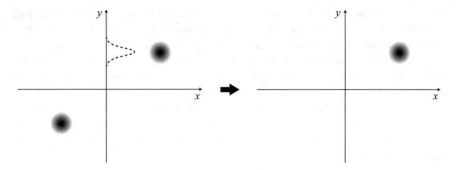

Fig. 5.4 Spontaneous collapse for an entangled system

process. In this way, the GRW theory justifies the folkloristic assumption that a measurement collapses the wave function of the measured system, although notably without attributing a special status to the measurement process or, worse, the "observer".

Finally, it is important to note that, if two particles (or systems) are in a product state, e.g.,

$$\psi(x, y) = \varphi_1(x)\varphi_2(y) = \frac{1}{N} \left\{ \exp\left[-\frac{(x + x_0)^2}{2\sigma^2}\right] + \exp\left[-\frac{(x - x_0)^2}{2\sigma^2}\right] \right\}$$
$$\times \left\{ \exp\left[-\frac{(y + y_0)^2}{2\sigma^2}\right] + \exp\left[-\frac{(y - y_0)^2}{2\sigma^2}\right] \right\},$$

the collapse of particle 2 will *not* affect the state of particle 1:

$$\psi_Y^2 \approx \frac{1}{N} \left\{ \exp\left[-\frac{(x + x_0)^2}{2\sigma^2}\right] + \exp\left[-\frac{(x - x_0)^2}{2\sigma^2}\right] \right\}$$
$$\times \left\{ \exp\left[-\frac{(2y_0)^2}{2a^2}\right] \exp\left[-\frac{(y + y_0)^2}{2\sigma^2}\right] + \exp\left[-\frac{(y - y_0)^2}{2\sigma^2}\right] \right\}$$
$$\approx \frac{1}{N} \left\{ \exp\left[-\frac{(x + x_0)^2}{2\sigma^2}\right] + \exp\left[-\frac{(x - x_0)^2}{2\sigma^2}\right] \right\} \exp\left[-\frac{(y - y_0)^2}{2\sigma^2}\right].$$

We can, for instance, think of the x-system as a subsystem and the y-system as its environment. We then see that, as long as we can consider the subsystem as isolated, that is, as long as we can assume that it's wave function factorizes, it will not be affected by collapse events in the environment. Moreover, the above computation is relevant to understanding why the quantum properties of quantum gases or Bose–Einstein condensates, in which the particles can be considered as independent, are not affected by spontaneous collapse.

Let's summarize. Spontaneous collapse is negligible for small (microscopic) systems—nuclear physicists can do business as usual—but wave functions describing a macroscopic superposition such as Schrödinger's cat are practically impossible. The surviving branches survive (to a very good approximation) with the quantum mechanical probabilities. The statistical analysis of measurement experiments and the appearance of "observable operators" in it then proceeds analogously to Bohmian mechanics or Many Worlds. This will be explained in detail in Chap. 7. Essentially, as long as the measurement problem is solved and pointer configurations follow Born statistics, the measurement formalism comes out the same in all quantum theories.

5.3   Remarks About Collapse Theories

A question that arises for collapse theories is this: what is the theory actually about, i.e., what is the *ontology* of the theory, and what exactly is the role of the wave function in it? The version of GRW presented here is really a theory about the flashes conceived as matter flashes. That is, a macroscopic object, like a chair, is "a galaxy" of such discrete events rather than a collection of persistent particles that move on continuous trajectories, and the role of the wave function is not to represent matter but to determine the probability of the random appearances of flashes. GRW with this flash ontology is called *GRWf*.

There are other ways to underpin collapse models with an ontology. In particular, we can use the wave function to define a *mass density*

$$m(\mathbf{x}, t) = \sum_{i=1}^{N} \int d^3q_1 \cdots d^3q_N \, \delta(\mathbf{x} - \mathbf{q}_i) |\psi(\mathbf{q}_1, \ldots, \mathbf{q}_N, t)|^2 \qquad (5.12)$$

that will behave reasonably in many situations (in particular, semi-classical ones) thanks to the collapse dynamics. This version of the theory is denoted by *GRMm*. Notably, the mass density exists in three-dimensional physical space, unlike the wave function, which is defined on the high-dimensional configuration space. GRWm thus postulates that material objects—in particular macro-objects like tables, chairs, cats, trees—correspond to configurations in a continuous matter field.

Many presentations of collapse theories do not involve any local ontology, however, but try to establish a connection with the physical world using the wave function alone. This is difficult to do, as we will discuss in more detail in Chap. 6 on Many Worlds. There is still a difference between a chair in physical space and the wave function of a chair—even a collapsed one—on the $3N$-dimensional configuration space.

Just like Bohmian mechanics (see Remark 4.6), collapse theories are nonlocal and therefore in accordance with Bell's theorem, as we shall discuss in Chap. 10. A collapse can instantaneously change the wave function and thus the distribution of flashes or mass densities throughout space. Another forward looking remark is that, although there exists a certain tension between nonlocality and Einsteinian relativity, collapse models—in particular with the flash ontology—can be generalized to relativistic spacetime without violating any principles of relativity. For Bohmian mechanics, this is not so simple. We will discuss all this in more detail in later chapters.

Finally, a basic difference between collapse theories on the one hand and Bohmian mechanics or Many Worlds on the other concerns the status of randomness and probabilities. In collapse theories, probabilities are part of the law and cannot be further derived or explained. In Bohmian mechanics (and arguably also Many Worlds), they are but typical appearances in a fundamentally deterministic theory. This does not amount to a difference in the phenomena, but very much in terms of what the laws of nature are fundamentally like.

The Many Worlds Theory

6

> Why doesn't our observer see a smeared out needle? The answer is quite simple. He behaves just as the apparatus did. When he looks at the needle (interacts), he himself becomes smeared out, but at the same time correlated to the apparatus, and hence to the system. [...] In other words, the observer himself has split into a number of observers, each of which sees a definite result of the measurement.
>
> Hugh Everett, *Probability in Wave Mechanics* (unpublished)

Hugh Everett III is credited as the father of the Many Worlds theory, although the name was only later introduced by Brice DeWitt, and it is disputed historically whether Everett really believed in the reality of many worlds. What is not disputed is Everett's insistence that we must take quantum mechanics seriously on all scales. He thus introduced the concept of the *universal wave function*, which has already played a crucial role in earlier chapters. Everett recognized that the shifty split, which the Copenhagen interpretation had introduced between the microscopic quantum regime and the macroscopic classical regime, couldn't be maintained if quantum mechanics was to provide a fundamental description of nature.

In contrast to David Bohm, however, Everett refused to introduce additional variables into the theory. He wanted to have a "pure wave mechanics", defined only in terms of the (universal) wave function and the linear Schrödinger equation. Today, it is generally accepted that such a theory results in a Many Worlds picture, in which various decoherent branches of the wave function describe equally real states. We, as observers, do not perceive such macroscopic superpositions because we take part in the branching ourselves. At the end of a spin measurement, let's say, there exists an electron with spin UP, a detector indicating spin UP, and an experimenter who sees a detector that indicates spin UP. And at the same time, there exists an electron with spin DOWN, a detector indicating spin DOWN, and an experimenter who sees a detector that indicates spin DOWN. Due to the linearity of the Schrödinger evolution, these two experimenters—or should we say, two copies

© Springer Nature Switzerland AG 2020
D. Dürr, D. Lazarovici, *Understanding Quantum Mechanics*,
https://doi.org/10.1007/978-3-030-40068-2_6

of the same experimenter?—cannot interact and directly perceive one another. They exist, as the story goes, in different worlds.

While the gist of Many Worlds comes out clearly in such scenarios, we shouldn't rely too much on the vague notion of "measurement" if we want to understand the theory as a fundamental description of nature. More generally, Everettian quantum mechanics, in its modern form, can be characterized by the following three principles:

1. At the fundamental level, the physical state of the universe is completely described by the universal wave function, whose evolution obeys the linear Schrödinger equation.
2. Under this time evolution, the universal wave function continuously splits into disjoint (decoherent) branches. This branching process is (for all practical purposes) irreversible, so that different branches will never overlap again in the relevant future.
3. Each branch of the universal wave function describes a macroscopically well-defined history in three-dimensional space, which we call a "world".

This gives rise to a physical description in which the universe—we stick to speaking about a single universe that comprises all of physical existence—exists in states in which we can identify a multitude of "worlds" or "histories" that are all simultaneously and equally real. In one such history, the cat is alive, jumps out of the box, and receives a bowl of milk. In another, the cat is dead, receives a funeral, and the experimenter gets sued by an animal rights group. Different such histories may have a common past but are causally disjoint with respect to their future evolution. The linearity of the Schrödinger equation ensures that different branches can never interact. And decoherence makes it practically impossible to bring macroscopic subsystems—such as the dead cat and the alive cat—into interference.

This is to say, in particular, that we have no direct empirical access to worlds other than our own. The reason for believing in their existence is simply that our best theory of nature predicts them—provided we accept the Many Worlds interpretation of quantum mechanics as our best theory of nature.

When we say that a world "splits" or "'branches" (for instance, in the course of a measurement experiment), we are actually talking about a gradual process. Think of a wave packet on an extremely high-dimensional configuration space fanning out into two or more parts that become more and more separated in that space. Don't try to think of an exact moment in which it goes "bing" and the world suddenly multiplies. The concept of a "world" has a certain vagueness—it's not possible, in general, to say exactly how many worlds exist or at what moment in time a new splitting has occurred. However, contrary to what the name might suggest, "worlds" are not fundamental in the Many Worlds theory. The fundamental description is always given by the wave function of the universe as a whole.

The idea of innumerable parallel worlds—many of which contain almost identical copies of ourself—is nonetheless bizarre. John Bell was more cautious and called it "extravagant", in particular "extravagantly vague".[1] In response, proponents of the Many Worlds theory like to point out that new physical theories have often led to radical changes in our understanding of nature, and forced us to accept that reality comprises much more than we could have guessed from everyday experience. The radical innovation of quantum mechanics, they say, is that the history of the universe is not a linear but a *branching* one.

6.1 Finding the World(s) in the Wave Function

On closer examination, the actual problem with the Many Worlds theory is not the many worlds, but the question of how to locate any world at all in the universal wave function. How can we relate the wave function of the universe on the extremely high-dimensional configuration space with the physical world that we experience in three-dimensional space? In other words, how do we find tables and chairs and cats and measurement devices with pointer positions in the universal wave function?

A first approximation to a solution goes as follows. Any point in $3N$-dimensional configuration space describes the positions of N particles in three-dimensional space, and this configuration can be such that it forms a table, or a cat, or a measurement device whose pointer points to the left. And other points in configuration space that lie nearby will describe particle configurations that deviate only slightly and will thus look macroscopically the same. In this way, we can identify entire regions of configuration space with certain macroscopic "images" that are coarse-grained from particle configurations. Thus, if a certain part of the wave function is (suitably well) localized in a region of configuration space that coarse-grains to a cat, we can say that we have located a cat in the wave function. And if another part of the wave function is (suitably well) localized in a region of configuration space that coarse-grains to a dead cat, we can say that we have located a dead cat in the wave function.

The problem with this explanation doesn't even lie in the term "suitably well" that the reader may find justifiably suspicious. The problem with this explanation is that we have been cheating all along. For what justifies the identification of points in $3N$-dimensional configuration space with configurations of N hypothetical particles in 3-dimensional space? What even justifies the name "configuration space" for the high-dimensional space on which the universal wave function lives? Configuration of what? If the ontology of quantum mechanics is supposed to be the wave function and the wave function alone, we cannot just pretend that its degrees of freedom refer, somehow, to particle positions.

In the modern literature, one thus finds another strategy that falls under the philosophical concept of "functionalism". The basic idea is the following. To be a cat is not to be a cat-shaped configuration of matter. To be a cat is to act like a

---

[1]J.S. Bell, *Speakable and Unspeakable in Quantum Mechanics*. Cambridge University Press, 2nd edn. 2004, p. 194.

cat: to chase after a mouse when it passes by, to purr when being caressed, to land on your feet when jumping out the window, etc. To locate a cat or a table or a chair in the universal wave function is thus not to find something that comprises a cat or a table or a chair (as Bohmian particles do), but to identify certain patterns in the wave function that, in their interplay, satisfy the causal and functional role of a cat or table or chair. Since the dynamics of quantum mechanics is in any case given by the wave function, we may expect (or hope) that the wave function will show the right dynamical behaviour to represent our world (and many others like it) in this way.

To make this more plausible, we recall our discussion of the classical limit, where we argued that the interactions of a system with its environment tend to create well-localized wave packets ("the environment constantly measures where the particles are") which propagate more or less like classical (Newtonian) bodies. Such wave packets thus form patterns in the universal wave functions that behave more or less like the macroscopic objects that we perceive in the world.[2]

If we think this through to the end, we'll see that this semi-classical behaviour is due to the Schrödinger equation that contains a second derivative with respect to "position" coordinates, and an interaction potential that is also given in terms of spatial coordinates, as if it described interactions in three-dimensional space. Many proponents of the Many Worlds theory would say that this form of the law is merely accidental, but given this law, we can run our functionalist analysis.

6.1.1   Everett Versus Bohm

It's questionable whether anyone could actually analyze a physical theory like that. However, if we grant that the Many Worlds theory can, at least in principle, represent our world (and many other worlds) in this way, we can also understand the main Everettian criticism of Bohmian mechanics. Recall Schrödinger's cat experiment. According to the Schrödinger equation, the wave function at the end of the experiment will always be a superposition of "dead cat" and "alive cat". In Bohmian mechanics, however, only one of these branches will guide the actual configuration of the system, corresponding to a dead cat *or* a living cat. The other wave packet—let's say of a dead cat, since the cat has actually survived—will be empty, and thanks to decoherence and the effective collapse, we can forget about it for all practical purposes. Of course, this doesn't mean that the empty wave packet has simply disappeared. It will continue to evolve according to the Schrödinger equation, and interact with the wave function of the experimenter and the laboratory and the rest of the universe, just as it would if it represented the actual state of the system. For an Everettian, however, the wave function of a dead cat is sufficient to say that there actually *is* a dead cat. She will thus insist that the other world with

---

[2]For a detailed discussion, see D. Wallace, *The Emergent Multiverse: Quantum Theory According to the Everett Interpretation*. Oxford University Press, 2012.

the dead cat (and the experimenter seeing a dead cat, and so on) exists even in the Bohmian theory. David Deutsch, one of the most vocal proponents of Many Worlds, thus proclaimed that Bohmian mechanics is nothing but a Many Worlds theory "in a state of chronic denial".

The Bohmian, of course, cannot accept this argument. For her, it is decidedly the particle configuration in three-dimensional space and not the wave function on the abstract configuration space that constitutes a world (or rather, *the* world). Instead, she will accuse the Everettian of not having local *beables* (in Bell's sense) in her theory, that is, the ontological variables that refer to localized entities in three-dimensional space or four-dimensional spacetime. The many worlds of her theory thus merely appear as a grotesque consequence of this omission.

The upshot is that we see a clash between two fundamentally different ways of describing nature: atomism versus monism, physical objects being composed of microscopic entities versus being functionally enacted by degrees of freedom in the wave function or quantum state. Many Worlds, however, is clearly the more radical and revisionist position, and we have to ask ourselves if there are sufficiently compelling arguments to take such a radical step. We will do that in Sect. 6.3. But first, we have to understand in what sense the Many Worlds theory even matches the phenomena.

6.2 Probabilities in the Many Worlds Theory

The Many Worlds theory has trouble reproducing the statistical predictions of quantum mechanics, i.e., Born's rule. The problem is not that the theory is deterministic (there is only one equation, the Schrödinger equation, which is deterministic). Since Boltzmann, it has been well understood how deterministic theories can ground statistical predictions. We have discussed this in detail in Chap. 3 and then applied Boltzmannian arguments to derive the Born rule in Bohmian mechanics. When it comes to probabilities in the Many Worlds theory, the critical question is rather: probabilities of what? The theory says, after all, that all possible results in a quantum experiment actually occur. There are thus no interesting probabilities in the sense of relative frequencies about which we could formulate a statistical hypothesis. All possible outcomes occur with probability 1.

If we return to our standard example of the spin-measurement on a spin-1/2 particle, it doesn't make sense to ask for the probability of measuring spin UP or spin DOWN: in one world, the upper detector clicks and we measure spin UP, in another world, the lower detector clicks and we measure spin DOWN (assuming, of course, that the particle was not in an eigenstate).

Naively, we may think that quantum statistics refer to the relative frequency of worlds. One outcome being "more likely" than another would then simply mean that it will be realized in a greater number of worlds. However, if this were true, the Many Worlds theory would actually make *incorrect* predictions. Suppose our "particle" is in the spin-state

$$\psi_1 = \frac{1}{\sqrt{2}}|\uparrow_z\rangle + \frac{1}{\sqrt{2}}|\downarrow_z\rangle$$

and we measure its spin in the z-direction. At the end of the experiment, our world will have split into two new branches: one in which we have measured spin UP, and one in which we have measured spin DOWN. Each possible outcome thus occurs in an equal number of worlds, in accordance with the quantum mechanical probabilities.

But now suppose the particle is instead in the spin-state

$$\psi_2 = \frac{1}{\sqrt{3}}|\uparrow_z\rangle + \sqrt{\frac{2}{3}}|\downarrow_z\rangle \,.$$

According to Born's rule, the probabilities are now 1/3 for spin UP and 2/3 for spin DOWN. According to the Many Worlds theory, however, the outcome is the same as before: we end up with two worlds, one in which the result of the measurement is spin UP and one in which it is spin DOWN, respectively. Hence, "counting worlds" (to the extent that this even makes sense) does *not* yield statistics that are consistent with the quantum mechanical predictions. And the "weights" of the world-branches, that is, the pre-factors $c_1 = 1/\sqrt{3}$ and $c_2 = \sqrt{2/3}$, don't have any immediate physical significance in the Many Worlds theory. It's not as though one world is "more real" or "exists with greater intensity" than the other. The functional and dynamical relations within a branch are all that matters.

Since it's difficult to find interesting statistical distributions in a Many Worlds universe, most authors try to locate probabilities in their heads, that is, they try to interpret them as *subjective probabilities*. For instance, after you perform a spin measurement—but before you look at the detector to see the result—you do not know whether you find yourself in a world in which the detector has registered "spin UP" or a world in which it has registered "spin DOWN". What should be your "degree of belief" in one or the other? If someone offers you a 2:1 bet on "spin UP", should you take it? The "chances" in this case arise from your "self-locating uncertainty"—you do not know which branch of the Many Worlds universe your present self inhabits—and the goal of any theoretical analysis would be to show that it's rational to assign your degrees of belief according to the quantum mechanical probabilities.

However, regardless of how convincing or unconvincing the proposed arguments may be, there is something unsatisfactory about retreating to purely subjective probabilities. After all, in our laboratories, we do not take bets or poll scientists on their personal expectations. We observe concrete statistical regularities in the world that can be reproduced in many independent experiments and are very well predicted by Born's rule. A quantum theory should be able to explain these empirical facts. Otherwise, the theory is no good.

6.2.1 Everett's Typicality Argument

Hugh Everett's own explanation of the Born rule (which, oddly, is endorsed only by a minority of modern Everettians) was based on a typicality argument—and thus on objective probability assignments—similar to the one we have discussed in Bohmian mechanics. Therein, the $|\Psi|^2$-measure defined by the universal wave function—or rather the absolute squares of the prefactors of the various branches of the universal wave function—define a typicality measure that is distinguished by the fact that it is stationary under the Schrödinger evolution. In the context of Many Worlds, this stationarity can be understood as follows: the "weight" associated with any branch at any time equals the sum of weights associated with all of its sub-branches at later times.

Consider for instance a sequence of z-spin measurements performed on identically prepared electrons in the state

$$\alpha|\uparrow_z\rangle + \beta|\downarrow_z\rangle, \quad |\alpha|^2 + |\beta|^2 = 1.$$

We denote by $|\Uparrow\rangle$ (respectively $|\Downarrow\rangle$) the state of the measurement device (and in the last instance the rest of the world) that has registered spin UP (respectively Spin DOWN). After the first measurement, our world splits according to

$$\alpha|\Uparrow\rangle|\uparrow_z\rangle_1 + \beta|\Downarrow\rangle|\downarrow_z\rangle_1, \tag{6.1}$$

where the index 1 indicates the first measurement. With the second measurement, each world splits anew, namely according to the decoherent wave branches:

$$\alpha^2|\Uparrow\Uparrow\rangle|\uparrow_z\rangle_2|\uparrow_z\rangle_1 + \beta\alpha|\Downarrow\Uparrow\rangle|\downarrow_z\rangle_2|\uparrow_z\rangle_1 + \alpha\beta|\Uparrow\Downarrow\rangle|\uparrow_z\rangle_2|\downarrow_z\rangle_1 + \beta^2|\Downarrow\Downarrow\rangle|\downarrow_z\rangle_2|\downarrow_z\rangle_1.$$

The first three steps of the branching are shown in Fig. 6.1. Conservation of the measure in each branch is now readily verified. For instance, in the left branch, after

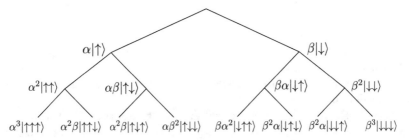

Fig. 6.1 Branching Many Worlds histories after three spin measurements. Graphic adapted from J.A. Barrett, Typicality in pure wave mechanics. Fluctuation and Noise Letters **15** (03), 1640009 (2016)

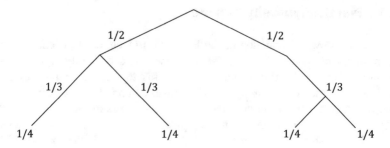

Fig. 6.2 Delayed measurements, all branches weighted equally. The measure is not conserved

the second measurement,[3] we have

$$|\alpha|^4 + |\alpha|^2|\beta|^2 = |\alpha|^2(|\alpha|^2 + |\beta|^2) = |\alpha|^2.$$

This conservation of the typicality measure wouldn't hold if we weighted each branch equally, i.e., if we simply counted the branches. This is easy to see if we assume that, in our example, the second measurements in the already separated worlds occur at different times. If the second measurement occurs earlier in the left branch than in the right branch, the total number of worlds increases from 2 to 3, and the measure of the right branch suddenly drops from $1/2$ to $1/3$. That is, until the second measurement occurs in the right branch as well, resulting in a total of four different worlds (Fig. 6.2).

Following the principle of stationarity, we thus arrive at the $|\Psi|^2$-measure as the typicality measure by which we weight branches of the wave function corresponding to Everettian worlds. And according to this measure, a *typical* branch will be one in which Born's rule—and thus quantum statistics—holds. We can check this for our example of consecutive spin measurements. After n measurements, the measure of worlds in which the outcome spin UP has occurred exactly k-times is

$$\binom{n}{k}|\alpha|^{2k}|\beta|^{2(n-k)}.$$

Writing $|\alpha|^2 =: p$ and $|\beta|^2 = 1 - p$, we see that this is a Bernoulli process with n trials and "probability of success" p. According to the law of large numbers, the *typical* relative frequencies for spin UP are thus $k/n \approx p = |\alpha|^2$, exactly matching the predictions of quantum mechanics.

In conclusion, Everett's analysis establishes that the Born statistics hold in *typical histories* of the constantly branching Many Worlds universe. We would now like to conclude the analysis with a solid probabilistic prediction and say something like:

---

[3]Spontaneous decoherence may cause the sub-branches to branch even further, but the total weight of branches registering a particular sequence of measurement results doesn't change.

"So, *I* should expect to experience a typical history in which the Born statistics hold." However, to whom this *I* actually refers in a Many Worlds universe is a difficult and maybe in the end philosophical question.

6.3 Many Worlds: A Brief Assessment

The Many Worlds theory is, all in all, a consistent way of thinking about quantum mechanics that is more serious than it may appear at first glance. So far, we have always avoided the term "interpretation", though in the context of the Many Worlds interpretation, it is perhaps appropriate. If quantum mechanics is supposed to be defined by the Schrödinger equation alone, then Many Worlds is not merely a possible interpretation, but the only honest one.

The key question, in the end, is whether there are good scientific reasons to accept such an extravagant and counterintuitive description of nature, if Bohmian mechanics and collapse theories (with caveats) offer empirically equivalent alternatives that are more down-to-earth, to say the least.

Many advocates of the Many Worlds theory take the economy of its mathematical formalism to be the decisive argument. The Many Worlds theory is defined by a single, relatively simple mathematical equation, the Schrödinger equation. Bohmian mechanics postulates, in addition, the guiding equation for the particle positions. Collapse theories can make do with a single equation, as well, but include additional parameters and stochastic terms that make the wave equation more cumbersome and less elegant than the linear Schrödinger equation. The conclusion that the Many Worlds theory offers the *simplest* explanation for quantum phenomena is, however, questionable, since Bohmian particles or the GRW flashes have an immediate connection with physical phenomena, while the Everettian has to tell some story about how patterns in the high-dimensional wave function relate to objects in the three-dimensional world that we experience. And this story is anything but simple—if it succeeds at all.

Another common argument is that the Many Worlds interpretation avoids the nonlocality of Bohmian mechanics or GRW that we have already mentioned. This claim is also questionable, though the issue is indeed more subtle than in other quantum theories where the nonlocality is evident. In Chap. 10, we will discuss Bell's theorem, which proves that certain statistical correlations predicted by quantum mechanics and observed in experiments cannot be explained by any local theory. According to the Many Worlds theory, however, these correlations occur only within individual branches, and a single branch does not comprise all of physical reality. Simply put, if measurements have no definite outcome, there is no worry about measurement outcomes being nonlocally influenced. On the other hand, it is still not possible, in general, to describe a localized system independently of other, arbitrarily remote systems. They only have a common state, given by the entangled wave function (which ensures that the right correlations manifest themselves in typical worlds). A compromise is often reached, according to which the Many Worlds theory has nonlocal states but no nonlocal interactions.

One possible point of view is that the fundamental stage for the Many Worlds theory is not three-dimensional space or four-dimensional spacetime, but the extremely high-dimensional configuration space (which is not actually a *configuration* space) in which the universal wave function lives as some kind of physical field. On this space, the theory is certainly local, but this has little to do with the sort of *spatio-temporal* locality that Einstein so deeply cared about. If the nonlocality of quantum mechanics appears bizarre or undesirable, it does so against the backdrop of certain intuitions about physical interactions between localized objects in three-dimensional space. It is not clear why a local description on a super-high-dimensional space, which is completely detached from our intuitions, should be less bizarre or undesirable.

More generally, we can conclude that Bohmian mechanics and GRW are theories about localized objects (particles or flashes, respectively), obeying a nonlocal law defined in terms of the wave function. In the Everettian theory, on the other hand, there is nothing but the wave function (on the fundamental level), which is itself a highly nonlocal object in a spatio-temporal sense.

One final (and related) argument is that Many Worlds is more compatible with relativity. In some sense, this is correct, since a relativistic generalization of the Many Worlds theory requires only a relativistic wave equation, while Bohmian mechanics or GRW require also a relativistic law for the particle motion, respectively the collapse dynamics. However, we don't even know yet how to formulate a consistent, relativistic, interacting theory on the level of the wave function (quantum field theory is mainly concerned with describing asymptotic scattering states). Therefore, a consistent, relativistic version of the Many Worlds theory does not yet exist. We will return to this difficult subject in Chap. 11.

The Measurement Process and Observables 7

Here are some words which, however legitimate and necessary in application, have no place in a *formulation* with any pretension to physical precision: *system, apparatus, environment, microscopic, macroscopic, reversible, irreversible, observable, information, measurement.* [...] On this list of bad words from good books, the worst of all is 'measurement'.

John S. Bell, Against Measurement[1]

When we look in textbooks on quantum theory to find the heart of the theory, one notion in particular stands out: the observable. In older representations, observables are self-adjoint operators on the Hilbert space of a system and they are supposed to describe the properties of a system which can be observed. In more modern texts (actually already since the 1950s), the notion of observables has been extended to *positive operator-valued measures* (POVMs). This extension was already in-between the lines in von Neumann's book *Mathematische Grundlagen der Quantenmechanik,*[2] because von Neumann viewed observables as emerging from their spectral decompositions, which are *projector-valued measures* (PVMs). All these seemingly abstract things will now be explained, and we shall see why they arose in the first place. We shall find that their role in quantum theory is that of bookkeepers for the statistics produced by measurement experiments. Once that is clear, we shall be immune to the philosophical encumbrances that come to light in Chap. 9 on hidden variables.

---

[1] In: J.S. Bell, *Speakable and Unspeakable in Quantum Mechanics* (2nd edn). Cambridge University Press, 2004, p. 214.

[2] J. von Neumann, *Mathematische Grundlagen der Quantenmechanik.* Springer, 1932.

© Springer Nature Switzerland AG 2020
D. Dürr, D. Lazarovici, *Understanding Quantum Mechanics*,
https://doi.org/10.1007/978-3-030-40068-2_7

7.1   Ideal Measurements: PVMs

The following mathematical analysis[3] is actually independent of the particular quantum theory one favours. However, considering ontological theories like Bohmian mechanics makes life considerably simpler because there is no need to introduce vague notions and concepts like "observer" or "collapse of the wave function". The mathematical analysis is essentially based on Born's statistical interpretation[4] (see Remark 1.2):

> If the wave function of a system is ψ, the particle positions of the system are $|\psi|^2$-distributed.

The various theories we have discussed may disagree about the meanings of terms like "particle positions" and "are distributed", but all theories will nevertheless agree in some way or other with Born's interpretation. In particular, there is agreement that Born's interpretation for the configurations of positions encompasses the empirical content of quantum mechanics. That may sound surprising, because "position" is in many textbooks merely introduced as one of many possible observables. But it should be remembered that the results of a measurement experiment are always associated with "pointer positions". The latter stand here as placeholders for the readouts of measurement results, which can be realised by a display or by a printout on a piece of paper or by a pointer pointing to a number on a scale. Whichever way they work, readouts are always expressed by configurations of matter in space. In this sense every experiment is at the end of the day a "measurement of positions". The statistics of measurement results corresponds to the statistics of pointer positions, that is, to configurations in space. Nothing more will be used in what follows.

Recall also the notation and the descriptions in Chap. 2, much of which we shall use here. As in the other chapters, we shall also make free use of the words "particle position" and "pointer position" as they are used in quantum mechanics. When we talk below about a "piece of apparatus", we take that term to mean possible pointer positions.

We consider, as we have already many times, an experiment $\mathscr{E}$, in which a system x (m-dimensional, represented by a wave function φ) is coupled to a piece of apparatus y (n-dimensional, represented by a wave function Φ—the ready state). The apparatus has discrete pointer positions corresponding to wave functions Φ_k, $k = 1, \ldots, N$, which can display the values $\{\alpha_1, \ldots, \alpha_N\} =: \mathscr{A}$. These correspond to the possible measurement results, and the associated wave functions

---

[3]D. Dürr, S. Goldstein and N. Zanghì, Quantum equilibrium and the role of operators as observables in quantum theory. In: *Quantum Physics Without Quantum Philosophy*, Springer, 2013.

[4]We use the following terminologies interchangeably: Born's statistical interpretation, Born's statistical law, Born's statistical rule. They all mean the same thing, although they sometimes emphasise different views.

are localized in configuration space, on the respective pointer positions indicating the result.

The Schrödinger evolution of the total wave function of the system and apparatus combined yields correlations between the pointer wave functions and a certain wave function of the system φ_k, because that is how the apparatus is intended to function:

$$\varphi_k(x)\Phi(y) \xrightarrow{\text{Schrödinger-evolution}} \varphi_k(x)\Phi_k(y). \tag{7.1}$$

Assuming that the wave functions are normalized, we obtain for the superposition

$$\varphi = \sum_k c_k \varphi_k, \quad \sum_k |c_k|^2 = 1, \tag{7.2}$$

$$\varphi(x)\Phi(y) \xrightarrow{\text{Schrödinger-evolution}} \sum_k c_k \varphi_k(x)\Phi_k(y), \tag{7.3}$$

which means that the initial wave function $\varphi = \sum c_k \varphi_k$ of the system becomes the new wave function φ_k with probability $|c_k|^2$.

Why is that? We already computed that in (2.4)–(2.6), but for a better appreciation, let us do it once again here, but rather briefly. According to Born's statistical law, the different pointer wave functions Φ_j must have disjoint supports, so the probability that the pointer position $Y \in \text{supp}\,\Phi_j$ is given by

$$\int\limits_{\{x,y|y\in\text{supp}\,\Phi_j\}} \left|\sum_k c_k \varphi_k(x)\Phi_k(y)\right|^2 \mathrm{d}^m x\, \mathrm{d}^n y \tag{7.4}$$

$$\stackrel{\text{supp}\,\Phi_k\cap\text{supp}\,\Phi_j\approx\emptyset}{=} |c_j|^2 \int |\varphi_j|^2(x)\mathrm{d}^m x \int |\Phi_j|^2(y)\mathrm{d}^n y = |c_j|^2,$$

where $\mathrm{d}^m x$ and $\mathrm{d}^n y$ are the corresponding higher-dimensional integration elements. If now $Y \in \text{supp}\,\Phi_j$, that is, if the pointer points let's say towards $\alpha_j \in \{\alpha_1, \ldots, \alpha_N\}$, then in each and every quantum theory it is the case that φ_j is the new (collapsed) wave function of the system. The other wave parts Φ_k, $k \neq j$, can for all practical purposes be ignored, either because of decoherence (loss of the ability to interfere) or because the collapse happened as a matter of fact (as in collapse theories, in which case they are gone forever). The right-hand side of (7.3) collapses to the product wave function $\varphi_j\Phi_j$, and as we discussed in Chap. 4 on Bohmian mechanics, we can regard φ_j as the new wave function of the system.

In short, the experiment $\mathscr{E}$ described by (7.1) provides us with a random linear map

$$\varphi \longrightarrow \varphi_j,$$

where φ_j occurs with probability $|c_j|^2$. We should be able to handle both the probabilities and the collapsed wave functions with ease, and one comfortable way of doing so is by observing the following simple consequence of (7.3). We recall that the Schrödinger evolution preserves the $|\psi|^2$-density. That is, in the language of Hilbert spaces the L^2-norm $\| \cdot \|$ is preserved, so

$$\|\varphi\Phi\|^2 \quad := \quad \int |\varphi(x)\Phi(y)|^2 \mathrm{d}^m x \mathrm{d}^n y$$

$$\overset{*}{=} \quad \int \left| \sum_k c_k \varphi_k(x)\Phi(y) \right|^2 \mathrm{d}^m x \mathrm{d}^n y$$

$$\overset{\text{Schrödinger evolution}}{=} \quad \int \left| \sum_k c_k \varphi_k(x)\Phi_k(y) \right|^2 \mathrm{d}^m x \mathrm{d}^n y$$

$$\overset{(7.4)}{=} \quad \sum_k |c_k|^2 \int |\varphi_k|^2(x) \int |\Phi_k|^2(y)\, \mathrm{d}^m x \mathrm{d}^n y$$

$$= \quad \sum_k |c_k|^2 . \tag{7.5}$$

By expanding and integrating the right-hand side of the equality $*$, we get

$$\sum_{k \neq k'} c_k^* c_{k'} \int \varphi_k^*(x)\varphi_{k'}(x)\mathrm{d}^m x = 0 . \tag{7.6}$$

Because that holds for arbitrary coefficients c_k, $c_{k'}$, each individual summand must be zero, that is,[5]

$$\int \varphi_k^*(x)\varphi_{k'}(x)\, \mathrm{d}^m x = 0 , \quad \text{for } k \neq k' . \tag{7.7}$$

This means that, for the measurement process (7.1) to happen, the system wave functions φ_k must form an *orthonormal system* with respect to the L^2-scalar product:

$$\langle \varphi | \psi \rangle := \int \varphi^*(x)\psi(x)\, \mathrm{d}^m x . \tag{7.8}$$

---

[5]For a quick proof, read (7.6) as a quadratic form $(c, Ac) = 0$ with vector $c = (c_1, \ldots, c_n) \in \mathbb{C}^n$ and the Hermitian matrix A with entries

$$A_{k,k'} := \int \varphi_k^*(x)\varphi_{k'}(x)\mathrm{d}^m x = A_{k',k}^+ .$$

Then, choosing the vectors c as eigenvectors, we see that all the matrix elements must vanish.

This should not come as a big surprise. The wave functions of the apparatus Φ_k and $\Phi_{k'}$ represent *macroscopically different pointer positions*, and hence they have disjoint supports in configuration space so they are already orthogonal. The φ_k inherit from that their own orthogonality, and of course the preservation of the L^2-norm by the Schrödinger evolution plays an essential role.

Hence, we can now look upon the representation (7.2) as a decomposition into an orthonormal system, and we can compute the coordinates c_k as usual by

$$c_k = \langle \varphi_k | \varphi \rangle = \int \varphi_k^*(x) \varphi(x) \, d^m x \,. \tag{7.9}$$

We also get the probability $|c_k|^2$ for the result α_k as

$$\mathbb{P}^\varphi(\alpha_k) = |c_k|^2 = |\langle \varphi_k | \varphi \rangle|^2 \,. \tag{7.10}$$

This leads to the introduction of *orthogonal projectors*

$$P_{\varphi_k} := \varphi_k \langle \varphi_k | \,, \tag{7.11}$$

with the agreement that the factor $\langle \varphi_k |$—a dual vector in the sense of linear algebra, which Paul Dirac called a *bra*-vector (see below)—acts upon wave functions ψ according to $\langle \varphi_k | \psi \rangle$. Using this we can write (7.2) in the form

$$\varphi = \sum_k \varphi_k \langle \varphi_k | \varphi \rangle = \sum_k P_{\varphi_k} \varphi \,. \tag{7.12}$$

The following properties of orthogonal projectors are easy to check:

$$P_{\varphi_k}^2 = P_{\varphi_k} P_{\varphi_k} = P_{\varphi_k} \qquad \text{(projector property)}, \tag{7.13a}$$

$$\langle P_{\varphi_k} \varphi | \psi \rangle = \langle \varphi | P_{\varphi_k} \psi \rangle \,, \quad \text{i.e., } P_{\varphi_k}^+ = P_{\varphi_k} \quad \text{(self-adjointness)} . \tag{7.13b}$$

The probability (7.10) of getting the value α_k can now be written as

$$\mathbb{P}^\varphi(\alpha_k) = \| P_{\varphi_k} \varphi \|^2 = \langle P_{\varphi_k} \varphi | P_{\varphi_k} \varphi \rangle = \langle \varphi | P_{\varphi_k} \varphi \rangle = |\langle \varphi_k | \varphi \rangle|^2 \,. \tag{7.14}$$

Remark 7.1 In Dirac's *bra-ket* notation, as it is known from standard quantum mechanics courses, we may write the projectors in the form $P_{\varphi_k} = |\varphi_k\rangle\langle\varphi_k|$, a powerful symbolism to which we shall return in Remark 7.4.

The moral is that some of the essential data of the experiment (7.1) are captured by the orthogonal projectors $(P_{\varphi_k})_k$. What about the measurement values? The set $\mathscr{A}$ of values should be captured as well, because statistical data like the mean or the variance of the α-values are of interest. This can be done compactly. The "operator"

$$\hat{A} = \sum_k \alpha_k P_{\varphi_k} \tag{7.15}$$

contains all we need. The probability of the value α_k is already given in (7.14). And when we compute the mean of the α-values, which we do in the usual probabilistic way by computing expected values, a nice expression turns up, namely,

$$\mathbb{E}^{\varphi}(\alpha) = \sum_k \alpha_k \mathbb{P}^{\varphi}(\alpha_k) = \sum_k \alpha_k \| P_{\varphi_k} \varphi \|^2$$

$$= \sum_k \alpha_k \langle \varphi | P_{\varphi_k} \varphi \rangle = \langle \varphi | \sum_k \alpha_k P_{\varphi_k} \varphi \rangle = \langle \varphi | \hat{A} \varphi \rangle . \tag{7.16}$$

Likewise for the "variance" of the α-values (a pleasant exercise for the reader!):

$$\mathbb{E}^{\varphi}(\alpha^2) = \sum_k \alpha_k^2 \mathbb{P}^{\varphi}(\alpha_k) = \langle \varphi | \sum_k \alpha_k^2 P_{\varphi_k} \varphi \rangle = \langle \varphi | \hat{A}^2 \varphi \rangle , \tag{7.17}$$

where we have used

$$P_{\varphi_k} P_{\varphi_{k'}} = 0 , \quad \text{for } k \neq k' ,$$

which follows because $(\varphi_k)_k$ is an orthonormal system. So the operator $\hat{A}$ turns out to contain all the relevant statistical information and, in their elegance, the right-hand sides seem to have a significance all of their own.

The operator $\hat{A}$ in (7.15) is associated with the experiment (7.1) by virtue of the corresponding family of orthogonal projectors and the measurement values (the real numbers on the display), and it is by construction a *self-adjoint operator*.[6] For a self-adjoint operator the family $(P_{\varphi_k})_k$ is uniquely defined (recall the diagonalisation procedure for Hermitian matrices in linear algebra courses) and it is called the *spectral decomposition* of the operator. In short, we go from experiment to the displayed values and projectors to the bookkeeping device, i.e., the operator:

$$\mathscr{E} \longmapsto (\alpha_k, P_{\varphi_k})_k \longmapsto \hat{A} . \tag{7.18}$$

In other words, the experiment defines the possible measurement values and the corresponding channeling into wave functions, which we read for convenience as eigenvalues and eigenvectors of a useful mathematical tool, the self-adjoint operator. The system's wave function which is fed into the experiment is not part of the

---

[6] A joker could also write imaginary numbers on the display of the apparatus. Then the bookkeeping device $\hat{A}$ with α_k imaginary would no longer be self-adjoint. One should reflect upon that.

bookkeeping device, since that can change, with the experimental setup itself held fixed.

What does this now have to do with projector-valued measures—the PVMs? This is simple! We just ask for the probability that a measured value lies in a subset $A \subset \mathcal{A}$. This is given in a purely logical way by adding up the individual probabilities, i.e., by virtue of (7.14), we have

$$\mathbb{P}^\varphi(A) = \sum_{k:\alpha_k \in A} \mathbb{P}^\varphi(\alpha_k) = \sum_{k:\alpha_k \in A} \langle \varphi | P_{\varphi_k} \varphi \rangle = \langle \varphi | \sum_{k:\alpha_k \in A} P_{\varphi_k} \varphi \rangle, \tag{7.19}$$

where

$$P_A := \sum_{k:\alpha_k \in A} P_{\varphi_k}$$

is again an orthogonal projector (the reader should check that). For obvious reasons, this map from sets of values A to orthogonal projectors P_A is called a projector-valued measure (PVM), because a measure is nothing but an additive set function, and it takes values in the space of projectors. Moreover, if we require the total probability to be equal to one, then with the identity operator Id, we must have

$$\sum_{k:\alpha_k \in \mathcal{A}} P_{\varphi_k} = \text{Id} .$$

We admit that the notion of measure seems a bit pretentious, because we are only dealing with discrete sets and counting is enough in this case. But later on we shall extend our description to the continuum, where the notion of measure is more appropriate.

In conclusion, we have here a measure (a PVM) on discrete subsets of values in $\mathcal{A}$. The PVM defines a self-adjoint operator—the bookkeeper $\hat{A}$—as in (7.18). That is how observables were introduced by von Neumann (see footnote 2).

Remark 7.2 (Example of Spin Measurement) An important example is the measurement of spin, which we already discussed in Chap. 1.7. We recall that, sending a wave function

$$\psi_0 = \phi_0(\mathbf{x}) \begin{pmatrix} \alpha \\ \beta \end{pmatrix}, \quad |\alpha|^2 + |\beta|^2 = 1 ,$$

with respect to the σ_z eigenbasis, through a Stern–Gerlach magnet oriented in the z-direction, the wave function splits into spatially separating parts

$$\psi_t = \phi_+(\mathbf{x}, t) \begin{pmatrix} \alpha \\ 0 \end{pmatrix} + \phi_-(\mathbf{x}, t) \begin{pmatrix} 0 \\ \beta \end{pmatrix} ,$$

where ϕ_+ and ϕ_- are normed to 1. The experiment is set up so that after a sufficient amount of time the wave parts will no longer overlap. The probabilities of spin UP and spin DOWN are now simply the probabilities that the particle will be found in ϕ_+ and ϕ_-, respectively. According to Born's statistical law, we compute

$$\mathbb{P}(\text{spin UP}) = \mathbb{P}(\mathbf{X} \in \text{supp}\,\phi_+) = |\alpha|^2 \int |\phi_+(\mathbf{x}, t)|^2 \mathrm{d}^3 x = |\alpha|^2 \,,$$

$$\mathbb{P}(\text{spin DOWN}) = \mathbb{P}(\mathbf{X} \in \text{supp}\,\phi_-) = |\beta|^2 \int |\phi_-(\mathbf{x}, t)|^2 \mathrm{d}^3 x = |\beta|^2 \,.$$

These probabilities can be read off easily from the projections onto the spin components $\binom{1}{0}$ and $\binom{0}{1}$, respectively. Written in the form of matrices, we have

$$P_+ = \begin{pmatrix} 1 & 0 \\ 0 & 0 \end{pmatrix}, \quad P_- = \begin{pmatrix} 0 & 0 \\ 0 & 1 \end{pmatrix},$$

and we obtain immediately

$$\langle \psi | P_+ \psi \rangle = |\alpha|^2 \,, \quad \langle \psi | P_- \psi \rangle = |\beta|^2 \,. \tag{7.20}$$

Omitting $\hbar$, the expected value is accordingly

$$\frac{1}{2}\mathbb{P}(\text{spin UP}) - \frac{1}{2}\mathbb{P}(\text{spin DOWN}) = \frac{1}{2}\langle \psi | P_+ \psi \rangle - \frac{1}{2}\langle \psi | P_- | \psi \rangle$$

$$= \langle \psi | \frac{1}{2}(P_+ - P_-)\psi \rangle = \langle \psi | \frac{1}{2}\sigma_z \psi \rangle \,.$$

Here the Pauli matrix $\sigma_z/2$ [see (1.27)] turns out to be the bookkeeping operator associated with the experiment. Generalising to an arbitrary spin measurement, say in the direction $\mathbf{a}$, we obtain $\hat{A} = \mathbf{a} \cdot \sigma/2$ as the associated operator, whose eigenvectors determine the PVM. In accordance with the usual *bra-ket* notation, we denote the latter by $|\uparrow_{\mathbf{a}}\rangle$ and $|\downarrow_{\mathbf{a}}\rangle$.

This particular example contains an interesting feature. We did not have to say anything about any measurement apparatus to get the orthogonality of the possible wave functions. That is already taken care of by the Schrödinger (actually Pauli) evolution of the system alone. At the end of the day we need only detect the wave part the particle is located in, and the wave function then collapses to the part corresponding to the observed outcome. This may lead to a deplorable way of speaking, referring to the matrix

$$\frac{1}{2}\mathbf{a} \cdot \sigma$$

as "the $\mathbf{a}$-spin observable", which can be "measured" by orienting a Stern–Gerlach magnet in the direction $\mathbf{a}$. Talking this way, we may fool ourselves into thinking that the spin observable represents a physical variable, and that we have really measured a predetermined value of a property of the system. This is not the case. Chapter 9 is about precisely this problem—the problem of hidden variables, which is really a "non problem", and so much so that the term "hidden variables" should be banned from the language of physics.

The situation is less simple for position or momentum (see Sect. 1.3), because in these cases there is no longer a discrete decomposition of the wave function corresponding to the possible measurement results. On the other hand the notion of measure becomes clearer. We shall come to this in a moment.

In general, the Hilbert space $\mathcal{H}$ of wave functions we need to consider will be infinite-dimensional, while the set of the α_k, the "measurement values", will be finite—in realistic experiments. Therefore, instead of one-dimensional orthogonal projectors, we shall generally have orthogonal projectors P_k onto the higher-dimensional subspaces $\mathcal{H}_{\alpha_k}$ containing those wave functions that correspond to the given pointer position pointing to α_k. Hence, in general,

$$\hat{A} = \sum_k \alpha_k P_k, \qquad \sum_k P_k = \mathrm{Id}. \tag{7.21}$$

Nothing much has changed. As above, we get bookkeeping operators from the projectors.

Now let us connect this to the relevant results from linear algebra. One topic of such a course is diagonalisation of Hermitian matrices, or more generally, self-adjoint operators. The spaces $\mathcal{H}_{\alpha_k}$ are the eigenspaces corresponding to the eigenvalues $\alpha_k \in \mathbb{R}$ of the self-adjoint operator $\hat{A}$. If the dimension of the eigenspace $\mathcal{H}_{\alpha_k}$ is greater than one, the eigenvalue α_k is said to be degenerate. The diagonalisation of a self-adjoint operator is achieved by the spectral theorem which is proven in a general setting in any course on functional analysis. We defined the operators via PVMs and the spectral theorem yields the converse, i.e., given a self-adjoint operator there corresponds a unique PVM, generally called the spectral decomposition in mathematics.

The spectral theorem is the reason for a certain independent standing acquired by observable operators in quantum mechanics (and it is the basis for the hidden variables question discussed in Chap. 9). It says that

$$\hat{A} = \sum_k \alpha_k P_{\alpha_k}, \quad \text{i.e., } \hat{A} \mapsto (\alpha_k, P_{\alpha_k}), \tag{7.22}$$

where it should be observed that, for the complete converse (including the physical meaning) of (7.18), the arrow to the experiment is missing. After all, would it make sense to write down an abstract (self-adjoint) operator and ask the experimenter to measure that?

Actually, it would be easy to forget where $\hat{A}$ came from in the first place, so the operator observables start having a life of their own, generating lots of confusion, because we are led to think that observables come first, and from them we get measurement values and eigenstates. So let us make an interim statement:

Remark 7.3 (Measurements of Observables) What does the utterance "measure an observable" actually mean? The observable is an abstract operator on a Hilbert space, i.e., it generalises the notion of matrix to infinitely many dimensions. For example, we consider the Laplace operator (multiplied by some dimensional factors) as the energy observable. So what does it mean to measure the Laplace operator? Well, its symbol is a triangle and someone with no further information might perhaps think they were supposed to measure its area, its perimeter, or some angles perhaps.

But what we actually mean is this: a measurement experiment $\mathscr{E}$ which is suitable for "measuring the observable $\hat{A}$" is one that produces a channeling as in (7.3), i.e., a splitting of the wave function into what can be read as the different eigenspaces of $\hat{A}$, and where the pointer points to one of the values $\alpha_1, \ldots, \alpha_n$, and the system wave function is collapsed (by whichever mechanism, see Chap. 2) onto the corresponding channel (eigenspace) with a probability to be computed from the corresponding projector by a formula like (7.14). That is what is expressed in (7.18).

Our discussion of measurement leading to the bookkeeping observables is highly idealized, and a whole hierarchy of generalisations of the notion of measurement exists. The type of experiment like (7.1) is also referred to as a "reproducible measurement". This means that, if in one run of the experiment, φ_j appears and the experiment is repeated with the initial wave function φ_j then the value α_j and the wave function φ_j will be obtained again with certainty. Note also that a measurement is said to be ideal if there is a unique wave function φ_j corresponding to the value α_j.

But there exist many non-reproducible measurements, for example those in which a detector absorbs the particle. Quite generally, in contrast with the ideal measurement, we may speak of a "formal measurement" of an observable, by which we mean every experiment which yields the eigenvalues with the corresponding quantum mechanical probabilities. The moral to be drawn is that the terminology "measurement of an observable" is tricky and necessarily so, until a clear relation is established with the physical process. It is very important to keep one thing in mind: a measurement is a process which in general *changes* the state of a quantum mechanical system. Many so-called quantum paradoxes are already resolved by taking that into account. An ideal measurement tells us more about the system *after* the measurement than about its state *before* the measurement, because we can conclude from the indicated value α_j that the system after the measurement (but not before) has wave function φ_j.

We may also consider types of measurements which essentially do not disturb the system. These will be considered in Chap. 8.

7.2 PVMs and POVMs in General

Here we say a little more about Born's statistical law $\rho^\varphi = |\varphi|^2$ and consider the abstract mathematical description of measurement experiments in general. This generality will include situations where the apparatus plays only a side role in the experiment, for example, when the apparatus is there simply to detect the part of the wave function in which the particle is evolving—as in a spin measurement. We associate a measurement experiment with a coarse-graining function

$$F : \mathscr{Q} \mapsto \mathscr{A} , \tag{7.23}$$

which is a random variable on the configuration space $\mathscr{Q}$ of all the particles involved in the experiment, and which focuses on the relevant measurement values $\mathscr{A}$. These can be values of the system coordinates x or values derived from the coordinates, like the configurations y which correspond to pointer positions.

Hence F maps the possible configurations q to the corresponding "measurement result" $F(q)$. In general, many (possibly infinitely many) configurations will correspond to the same measurement result. This is clear if we think of the measurement apparatus, where very many micro-configurations represent the same pointer position. This is why F is a coarse-graining variable.

Once again, we split the general configuration $q \in \mathscr{Q}$ into $q = (x, y)$. The most general structure we can deduce from the $|\varphi|^2$-statistic results from the following sequence:

$$\varphi(x) \quad \xrightarrow[\text{to apparatus}]{\text{system couples (possibly)}} \quad \Psi(x, y) = \varphi(x)\Phi(y)$$

$$\xrightarrow[\text{of system and apparatus}]{\text{Schrödinger evolution}} \quad \Psi_T(x, y)$$

$$\xLongrightarrow{\text{Born's statistical law}} \quad \rho^{\Psi_T} = |\Psi_T|^2(x, y) \tag{7.24}$$

$$\xrightarrow{\text{we are only interested in}} \quad \mathbb{P}^\varphi(d\alpha) = \mathbb{P}^{\Psi_T}\left(F^{-1}(d\alpha)\right) . \tag{7.25}$$

The single arrows represent linear maps. The first signifies multiplication by the wave function of the apparatus and the second the linear evolution according to the Schrödinger equation. The "possible coupling" above the first arrow indicates that we also wish to consider situations in which the apparatus does not play any significant role and in which we only wish to study the statistics of the system coordinates. T is in general a long time, the time at which the experiment ends. The third arrow now signifies[7] Born's statistical law, while the last arrow indicates

---

[7] Actually a sesquilinear form, linear in the first argument and antilinear in the second.

that we compute the relevant probabilities with the quadratic form: $\mathbb{P}^\varphi(d\alpha)$ is then Born's probability that the displayed value α lies in the (infinitesimal) value set $d\alpha \subset \mathscr{A}$. The right-hand side of (7.25) is nothing but

$$\int_{F^{-1}(d\alpha)} |\Psi_T(x, y)|^2 \, d^m x \, d^n y \,,$$

where the pre-image set $F^{-1}(d\alpha)$ contains exactly those configurations which are coarse-grained to the value range $d\alpha$.

The total wave function Ψ_T thus depends linearly on the system wave function φ, and the probabilities (7.25) depend sesquilinearly (like a unitary scalar product) on Ψ_T. Altogether the sequence represents a sesquilinear map of the system wave function φ to the probabilities, i.e., a map of the form

$$\varphi \mapsto \mathbb{P}^\varphi(d\alpha) =: \langle \varphi, O(d\alpha)\varphi \rangle \geq 0 \,, \tag{7.26}$$

where the right-hand expression is nothing but a sesquilinear form on φ. It is a general theorem of functional analysis that a positive-definite sesquilinear form can be expressed as a scalar product with a positive operator, just as the right-hand side of (7.26) already shows. In the present case, the positive operators are indexed by sets, for example, by the "infinitesimal set" $d\alpha$. Hence, for a general set $A \subset \mathscr{A}$, we have

$$O(A) = \int_A O(d\alpha) \,, \quad \mathbb{P}^\varphi(A) = \langle \varphi, O(A)\varphi \rangle = \int_A \langle \varphi, O(d\alpha)\varphi \rangle \,.$$

The family of operators constructed in this way thus yields a *positive operator-valued measure*, or POVM. In particular, the following properties are naturally fulfilled:

$$O(\mathscr{A}) = \int_{\mathscr{A}} O(d\alpha) = \mathbb{P}^{\Psi_T}\left(F^{-1}(\mathscr{A})\right) = \mathbb{P}^{\Psi_T}(\mathscr{Q}) = 1 \tag{7.27}$$

and

$$O(A \cup B) = O(A) + O(B) \,, \quad \text{for } A, B \subset \mathscr{A} \text{ disjoint.} \tag{7.28}$$

These are consequences of Born's statistical law, which we accept without ifs and buts. Therefore, the following theorem holds[8]:

Theorem 7.1 (Justification of the Operator Formalism) *The statistics of measured values (or displays of pointers) are in principle given by POVMs.*

---

[8] See, however, Remark 7.5.

Let us now discuss some examples.

First we consider the imbedding of the already discussed PVMs, since they are special cases of POVMs in which the positive operators are orthogonal projectors. In (7.19), we have a discrete measure on the value set $\mathscr{A}$ (more precisely, on its power set). In accordance with (7.25), this is now represented by

$$
\begin{aligned}
\mathbb{P}^{\varphi}(\{\alpha_{k_1}, \ldots, \alpha_{k_n}\}) &= \mathbb{P}^{\Psi_T}\left(F^{-1}(\{\alpha_{k_1}, \ldots, \alpha_{k_n}\})\right) \\
&= \mathbb{P}^{\Psi_T}\left(\{(\mathbf{x}, \mathbf{y}) | F(\mathbf{x}, \mathbf{y}) \in \{\alpha_{k_1}, \ldots, \alpha_{k_n}\}\}\right) \\
&= \sum_i \|P_{\varphi_{k_i}}\varphi\|^2 = \sum_i \langle \varphi | P_{\varphi_{k_i}}\varphi \rangle = \sum_i \langle \varphi | \varphi_{k_i} \rangle \langle \varphi_{k_i} | \varphi \rangle \\
&= \langle \varphi | \sum_i P_{\varphi_{k_i}}\varphi \rangle .
\end{aligned}
\tag{7.29}
$$

In the case of Stern–Gerlach spin experiments (see Remark 7.2), no coupling to apparatus is required. It suffices to send the wave function through the Stern–Gerlach magnet. T is the time at which the wave function has left the magnet. The coarse-graining variable is then $F(\mathbf{x}, \mathbf{y}) = F(\mathbf{x}) \in \{-1/2, 1/2\}$, while the probability is

$$
\mathbb{P}^{\varphi}\left(\left\{F^{-1}\left(\pm\frac{1}{2}\right)\right\}\right) = \left\| P_{\pm} \binom{\psi_1}{\psi_2} \right\|^2 .
$$

This yields a PVM as well.

An example of a continuous value set—but also without coupling to apparatus—is provided by the statistics of the position $\mathbf{x} \in \mathbb{R}^3$ of a particle. The map F is now simply the identity, that is, the measured values are $\alpha = F(\mathbf{x}) = \mathbf{x}$, while $\mathscr{A}$ is the configuration space of the system and $\Psi_T = \varphi$. What remains is the quadratic form

$$
\varphi(\mathbf{x}) \Longrightarrow \rho^{\varphi}(\mathbf{x})
$$

and the probability $\mathbb{P}^{\varphi}(\mathrm{d}^3 x) = \rho^{\varphi}(\mathbf{x})\mathrm{d}^3 x$. Hence, we obtain

$$
\mathbb{P}^{\varphi}(A) = \int 1_A |\varphi|^2 \, \mathrm{d}^3 x = \int_A \langle \varphi | 1_{\{\mathrm{d}^3 x\}} | \varphi \rangle = \langle \varphi | 1_A | \varphi \rangle ,
\tag{7.30}
$$

where 1_A is the characteristic function of the set A. Here, $O(\mathrm{d} x^3) := 1_{\{\mathrm{d}^3 x\}}$ acts like P_{φ_k}, that is, like a projector, but now, taking the element $\mathrm{d}^3 x$ seriously, it is not a projector onto wave functions but rather a projector-valued measure.

Once again, in the same way that the $(P_{\varphi_k})_k$ form a family of orthogonal projectors indexed by the values k (as placeholders for α_k), we now see that $(O(A))_{A\subset\mathbb{R}}$ is a family of orthogonal projectors. Because of the continuous set of values, it is indexed by (measurable[9]) subsets of the set of all values. The projectors are thus defined by

$$O(A) : \varphi \longmapsto \mathbb{1}_A \varphi(\mathbf{x}) = \begin{cases} \varphi(\mathbf{x}) \,, & \mathbf{x} \in A \,, \\ 0 \,, & \text{otherwise} \,. \end{cases} \tag{7.31}$$

This now makes the notion of *operator-valued measure* very clear: the measure evaluates subsets of the set of values relevant to us—the image set of F. But in contrast to the measures used in analysis courses, like the Lebesgue measure, this measure takes values in the set of operators acting on wave functions.

The examples given so far are PVMs. The projector property has been shown in the discrete setting. For the position PVM, we can quickly show the properties as well (good exercise):

$$O(A)^2 = O(A) \,, \quad O(A)O(B) = O(A \cap B) \,,$$

so that, for $A \cap B = \emptyset$, the orthogonality is obvious, with the agreement that $O(\emptyset) = 0$. As in the discrete case, the PVM can be associated with a self-adjoint operator as bookkeeper. Here, it is the *position observable*

$$\hat{\mathbf{X}} := \int_{\mathbb{R}^3} \mathbf{x}\, \mathbb{1}_{\{d^3 x\}} \,.$$

Recalling (4.32), we also have an example of a nontrivial F mapping to a continuous set of values. For the statistics of the momenta (actually, the asymptotic velocities), we would choose

$$F(\mathbf{x}) = \frac{\mathbf{X}(\mathbf{x}, T) - \mathbf{x}}{T} \,,$$

for large T. Recalling also (1.11), for large times, the wave function is channeled into a continuum of plane waves $\varphi_k \sim e^{i\mathbf{k}\cdot\mathbf{x}}$. The corresponding PVM is perhaps a bit clumsy to write down. It looks much more elegant in the Dirac notation.

Remark 7.4 (On the Dirac Notation)   We recall the right-hand side of (7.14). This is analogous to the right-hand side of (7.30) which describes the distribution for position measurements. In the Dirac formalism (see Remark 7.1) the right-hand

---

[9]In the sense of measure theory.

side of (7.30) is written as

$$\mathbb{P}^\varphi(A) = \int_A \langle\varphi|\mathbb{1}_{\{d^3 x\}}|\varphi\rangle = \int_A \langle\varphi|\mathbf{x}\rangle\langle\mathbf{x}|\varphi\rangle d^3 x \,,$$

so the analogy between (7.29) and (7.30) becomes even clearer. Note that $|\mathbf{x}\rangle$ is not a wave function, but it is conventional to express $\langle\mathbf{x}|\varphi\rangle := \varphi(\mathbf{x})$. In analogy with (7.21), it is usual to write $|\mathbf{x}\rangle\langle\mathbf{x}| d^3 x$ for $\mathbb{1}_{\{d^3 x\}}$, whence the position observable can be written as

$$\hat{\mathbf{X}} = \int_{\mathbb{R}^3} \mathbf{x}|\mathbf{x}\rangle\langle\mathbf{x}| d^3 x \,.$$

It serves little purpose to understand $|\mathbf{x}\rangle$ in some abstract mathematical sense as a wave function. But morally, we can do so if we imagine that the particle is with certainty at the position $\mathbf{x}$. In fact, the product $\langle\mathbf{x}|\mathbf{x}'\rangle = \delta(\mathbf{x} - \mathbf{x}')$ yields the δ-function. The Dirac formalism is a powerful symbolism which gets its mathematical basis from PVMs. However, it does not mean any more than what is expressed in (7.29) and (7.30).

By analogy, we can now reexpress the momentum observable which we derived in Sect. 1.3. The PVM is in this case $|\mathbf{k}\rangle\langle\mathbf{k}| d^3 k$, where the action of the PVM $|\mathbf{k}\rangle\langle\mathbf{k}| d^3 k$ on a wave function ψ is defined by

$$\langle\mathbf{x}|\mathbf{k}\rangle\langle\mathbf{k}|\psi\rangle d^3 k = \frac{1}{\sqrt{2\pi}^3} e^{i\mathbf{k}\cdot\mathbf{x}} \hat{\psi}(\mathbf{k}) \, d^3 k \,.$$

The momentum observable is thus

$$\hat{\mathbf{P}} = \int_{\mathbb{R}^3} \mathbf{k}|\mathbf{k}\rangle\langle\mathbf{k}| d^3 k \,.$$

As we have already said, the relevant structure that emerges from the sequence (7.24) is more general. Instead of projector-valued measures (PVMs), we have in general genuine positive operator-valued measures (POVMs), which are not to be put in correspondence with self-adjoint operators as observables. A nice example, thanks to its particular relevance, is this. We transform our position PVM into a true POVM by thinking of a detector which is characterized by an intrinsic measurement error. This means that the measurement results are contaminated by an error which arises from the apparatus itself. To do this, let $p(\mathbf{x})$ be a probability density on $\mathbb{R}^3$, modelling the uncertainty due to the apparatus. The measured position value $\tilde{\mathbf{X}}$ can thus be expressed in the form $\mathbf{X} + \mathbf{Y}$, the $|\varphi|^2$-distributed position $\mathbf{X}$ plus the p-distributed measurement error $\mathbf{Y}$. In general, we may assume that $\mathbf{X}$ and $\mathbf{Y}$ are independent, i.e., the distribution of $\tilde{\mathbf{X}}$ is given by the convolution[10]

$$\tilde{\rho}(\mathbf{x}) = \int p(\mathbf{x} - \mathbf{y})|\varphi|^2(\mathbf{y})\mathrm{d}^3 y\,.$$

Hence, the probability is

$$\mathbb{P}^{\varphi}(\tilde{\mathbf{X}} \in A) = \int_A \tilde{\rho}(\mathbf{x})\mathrm{d}^3 x = \int_A \int p(\mathbf{x} - \mathbf{y})|\varphi|^2(\mathbf{y})\mathrm{d}^n y\,\mathrm{d}^n x$$

$$= \int \left[\int \mathbb{1}_A(\mathbf{x}) p(\mathbf{x} - \mathbf{y})\mathrm{d}^3 x \right] |\varphi|^2(\mathbf{y})\mathrm{d}^3 y$$

$$=: \langle \varphi | \tilde{O}(A)\varphi \rangle\,. \tag{7.32}$$

This is straightforward probability calculus. The multiplication operator

$$\tilde{O}(A) = \int p(\mathbf{x} - \mathbf{y})\mathbb{1}_A(\mathbf{y})\mathrm{d}^3 y\,,$$

that is,

$$\tilde{O}(A) : \varphi \longmapsto \int_A p(\mathbf{x} - \mathbf{y})\mathrm{d}^3 y \varphi(\mathbf{x})\,, \tag{7.33}$$

now defines a true POVM (the reader should check this)

$$\tilde{O}(A)^2 \neq \tilde{O}(A)\,,$$

where the equality holds only for $p(\mathbf{x}) = \delta(\mathbf{x})$, in which case the POVM becomes the position PVM (7.31). If we are interested in the variance of the variable $\tilde{X}$, we calculate that as usual, considering $x \in \mathbb{R}$ for simplicity:

$$\mathbb{E}(\tilde{X}^2) = \int x^2 \tilde{\rho}(x)\,\mathrm{d}x = \int x^2 \langle \varphi | \tilde{O}(\{\mathrm{d}x\})\varphi \rangle$$

$$= \int x^2 p(x - y)|\varphi|^2(y)\,\mathrm{d}y\,\mathrm{d}x\,. \tag{7.34}$$

Not much more needs to be said, except that this should be compared with (7.17): it simply makes no sense to associate a bookkeeping observable $\hat{A}$ with an experiment

---

[10]Consider the Fourier transform

$$\hat{\tilde{\rho}} = \mathbb{E}(\mathrm{e}^{\mathrm{i}\alpha \cdot \tilde{\mathbf{X}}}) = \mathbb{E}(\mathrm{e}^{\mathrm{i}\alpha \cdot (\mathbf{X} + \mathbf{Y})}) = \mathbb{E}(\mathrm{e}^{\mathrm{i}\alpha \cdot \mathbf{X}})\mathbb{E}(\mathrm{e}^{\mathrm{i}\alpha \cdot \mathbf{Y}}) = \widehat{|\varphi|^2} \cdot \hat{p}\,,$$

and note that the Fourier transform of a product is a convolution, as is easily checked.

which is described by a true POVM, because that does not bring any computational advantage of the kind in (7.17). It is really the POVM which encodes all the relevant details. A further POVM will be discussed in Remark 9.1.

Remark 7.5 (On the Importance of Being a POVM)   POVMs describe the statistics of measurement experiments, yielding answers for arbitrary wave functions φ. That is, POVMs are abstract operators on the entire Hilbert space of the system. But in real experiments, we cannot prepare arbitrary wave functions and we have at best access to only a few special wave functions. An example of this are experiments to make time measurements in quantum mechanics, such as tunnelling times or the arrival times of particles.[11] In this sense the discussed abstract structures are mathematical superstructures which do not really contribute to everyday physics. At the end of the day, all that counts is Born's $\rho = |\psi|^2$.

7.3   It Is Theory that Decides What Is Observable

What we say in the following has to do with the notion of "physical reality".[12] It shouldn't be necessary to waste too many words on it (at least not in a book that is primarily about physics), but some, at least, may be helpful.

Through our senses, we experience first and foremost a reality that is, so to speak, inside us—in our mind or consciousness. But physics is not about this internal experience—for *whose* experience should that be? Physics is necessarily about the external world which is common to all (or at least conceived as such). It is, for instance, about the moon we perceive, not about our perceptions of the moon. This insight goes back a long way. In his essay *Nature and the Greeks*, Erwin Schrödinger identifies the great pre-Socratic philosopher Heraclitus (circa 500 BC) as one of the first to separate private experience from the lawful cosmos that is the same for all of us:

> The waking have one common world, but the sleeping turn aside each into a world of their own.
>
> Heraclitus, Fragment DK B 89[13]

There must then be an obvious connection between the world we perceive and the world described by the physical theory. This is clearly the case for theories

---

[11]For a recent such proposal, see Siddhant Das and Detlef Dürr, Arrival time distributions of spin-1/2 particles, Scientific Reports **9**, 1–8 (2019).

[12]The section title here is taken from a remark made by Einstein to Heisenberg, quoted from *Physics and Beyond: Encounters and Conversations*. A.J. Pomerans, trans. (Harper and Row, New York, 1971), Chap. 5.

[13]Translated by the authors from Bruno Snell (Hg.), *Fragmente: Griechisch - Deutsch (Sammlung Tusculum)*. Artemis & Winkler, 2007, p. 29.

like Bohmian mechanics or GRW that postulate an ontology of localized objects in three-dimensional space—particles or flashes that can make up a table or a moon or the pointer of a measurement device—but less clear for theories like Many Worlds that describe first and foremost a wave function or "quantum state". In any case, a theory must make predictions that are accordance with our experience. It must be *empirically adequate*, or else the theory is no good. The following point, however, is often ignored: we can only talk about physical reality in terms of a physical theory.

This leads to a question about the "observability" of structures and variables that enter the formulation of a theory. The Copenhagen approach to quantum mechanics gave rise to a rather pernicious form of the question: Are *all* the variables which enter your theory observable?—the implication being that a physical theory should *only* contain observable quantities, like for example the operator observables, which can be "measured". (But beware! Read Remark 7.3.) Perhaps it is somewhat amusing to remark in this context that the wave function is not measurable in the sense of the quantum mechanical measurement process discussed in this chapter. This is a direct consequence of the superposition principle [see (2.3)]. For think of a piece of apparatus that has wave functions on its display, i.e., instead of digits 1 and 2, or "cat alive" and "cat dead", we see the display $\varphi_1, \ldots, \varphi_n$. Then (2.3) shows after just a little thought that no such apparatus could actually exist. In the Copenhagen interpretation, this is necessarily ignored.

In any case, the precise quantum theories "without observers" which we present in this book do not satisfy the requirement of only dealing with "measurable" quantities. (And why should they? Some variables play a role in describing or predicting measurements—and, more generally, in describing or predicting what the world is like—without being "measurable" themselves.) In the Many Worlds theory, this is obvious, not only because we have no empirical access to the other "worlds" but also because of the previous remark that the wave function itself is not measurable.

Perhaps surprisingly, the question is asked most insistently in the context of Bohmian mechanics: Can we measure the Bohmian trajectories or the Bohmian velocity of a particle? This is supposed to mean: Can we design a piece of apparatus which measures and displays the trajectory taken by the Bohmian particle or which measures and displays the velocity?

On the one hand it is clear that some Bohmian trajectories can be measured: the trajectory of a particle in a cloud chamber is a Bohmian trajectory, and so is the path on which the pointer of a measurement device moves. But these are situations that involve many particles and that can be described in more or less Newtonian terms. It is not what people have in mind they ask the question. What they would actually like to measure are the microscopic trajectory of a particle which is guided by its wave function.

For instance, in the double slit experiment, can we see the particle moving on its trajectory through one of the slits and still obtain the interference pattern? The answer is clear: according to Born's statistical law, a position measurement changes the wave function and hence also the Bohmian trajectories. More precisely, if we observe the particle, its wave function entangles with the observing system and the relevant wave function is then one which lives on the configuration space of all the particles which are involved, including those making up the apparatus. This will in general destroy the interference, since decoherence takes over. We have already talked about that, so there's nothing new here really. We can observe the particle's motion either at the slit or at the screen, but we cannot observe the trajectories responsible for the interference pattern.

What should we conclude from that? Some would say that *if we cannot observe the particle trajectories, then they do not exist!* Such a conclusion is obviously nonsense: the Bohmian theory describes a world of particle trajectories and this world could in fact be ours, because the observable phenomena that the theory predicts are in accordance with the phenomena that we do in fact experience.

Others would say that *if we cannot observe the particle trajectories, then we may just as well omit them from the theory.* That is nonsense as well, because there must be something that is actually "recorded" by the black spots on the screen and, more basically, something that the screen, and the laboratory, and the experimenter are made of in the first place. In Bohmian mechanics, that something is a configuration of particles which do have positions and thus a trajectory. In other theories, that might perhaps be flashes or matter fields, but certainly not observable operators and most certainly not "nothing". More generally, a theory without a clear ontology is no theory at all.

Alright, so what about the idea that there are particles and positions of particles, but only when we observe them? That is the worst nonsense of all, and all that needs to be said about that was said in Chap. 2.

Macroscopic objects do have a well defined position (at least that is how we experience them) and the simplest way to explain that is that they consist of microscopic objects which themselves have definite positions at all times. It is true that the trajectories of macroscopic objects can in general be measured without disturbing them too much, but not so for microscopic objects. We should be able to explain that, and Bohmian mechanics does so. In the GRW theory, for example, the microscopic objects do not move on continuous paths. Here we need to explain why it appears to be the case that macroscopic objects move on continuous trajectories, and the GRW theory does so.

So what is really out there in the universe? Are there particles, flashes, strings, or something else entirely? That is indeed a good and difficult question, which cannot be decided solely on empirical grounds. This is why we engage categories like "beauty" or "simplicity" which may help us to decide. It's okay to follow the feeling that we live in a beautiful rational cosmos, and that is how we should perceive the theory: simple, beautiful, comprehensible—so it feels just right.

But leaving aside the question of the "true" description of Nature, the message is that a physical theory is our way to explain the world, to speak about and make sense of the physical universe. In other words, there is without question a world out there, but the way we conceive of it, nay, what the world is like, that is given to us by the theory. And that is why it is the theory that tells us what a measurement is and what it is that can be measured! Without Maxwell's theory of electromagnetism, we would never measure electromagnetic fields. Of course, we would see "light"—for example, when the Sun lights up the day—and we may perhaps try despairingly to find an explanation for a sunburn, but that would not count as an observation of electromagnetic fields.

We shall end now with the precise answer to the question as to whether we can measure Bohmian velocities. The answer is *not by a measurement in the sense of the sequence (7.24)*. That sequence has the consequence that the displayed values are distributed according to [see, for example, (7.29)]

$$\mathbb{P}^{\psi}(A) = \text{Bilinear form}[\psi](A)$$

on the subsets A of the value set when the initial wave function is ψ. The relevant observation is that this is a bilinear form. Like binomial formulas, this always satisfies

$$\mathbb{P}^{\psi_1+\psi_2}(A) \le 2\mathbb{P}^{\psi_1}(A) + 2\mathbb{P}^{\psi_2}(A), \tag{7.35}$$

and we shall put this to good use. Consider the velocity

$$\mathbf{v}^{\psi} = \frac{\hbar}{m}\text{Im}\frac{\nabla\psi(\mathbf{Q}, t)}{\psi(\mathbf{Q}, t)}$$

of the Bohmian particle. A complex superposition of two real wave functions yields a complex-valued wave function and this generates in general a nonzero velocity vector field. But the vector field generated by real wave functions is zero everywhere, and this contradicts the relation (7.35) if A is a subset which does not contain zero. Therefore we cannot measure the velocity.

There is nevertheless a way to know the wave function and to measure the trajectories in a certain sense, by drawing more heavily on theoretical inferences. For example, take the ground state wave function of the electron in a hydrogen atom. That wave function is real and the Bohmian velocity is thus zero. If the atom is prepared at time T_1 and we measure the position of the electron (possible at least in principle) at time T_2, we know that the trajectory was at that position for the entire time interval $[T_1, T_2]$.

In the double slit experiment, in a similar manner, we can infer from the theory the slit the particle went through, at least in idealized situations. Assuming a symmetric setup, we know that if the particle hits the screen above the symmetry axis, it moved through the upper slit, and if it hits the screen below the axis, it moved through the lower slit. This is because, for each trajectory crossing the symmetry line from above, there would be one which crosses from below. But different Bohmian trajectories cannot intersect, being the unique solutions of a first order differential equation.

It is thus important to be clear about what we mean by "measurements" and why we care about them in the first place. At the end of the day, the notion of measurement is not such a simple one. As scientists, we want to find out things about the world, if necessary by measuring and quantifying them. But we don't construct theories to predict experiments. We perform experiments to test and inform theories, which are our means of understanding the world. At the same time, the theory plays a crucial role in devising and interpreting experiments. That is why we said above that Bohmian velocities cannot be measured *by a measurement in the sense of the sequence* (7.24).

In fact, in recent years, an extended notion of measurement has been receiving more and more attention, and it has become possible to use a series of such measurements to reconstruct the Bohmian trajectories after they have passed the double slit. These are the so-called *weak measurements*, which we shall discuss next.

Weak Measurements of Trajectories

8

> However, as a matter of principle, it is quite wrong to want to base a theory only on observable quantities. Because in reality it is the other way around. It is theory that decides what is observable.
>
> Albert Einstein, as quoted by Werner Heisenberg[1]

We showed in the previous chapter that the Bohmian velocity of a particle cannot be measured according to (7.24). But by now there are many measurement experiments which report on successful measurements of Bohmian velocities. So what is going on? The answer is that these measurements are not of the form so far discussed. The new way of measuring is called *weak measurement*. It is a measurement in which the wave function of the measured particle (system) is only weakly disturbed. The theory of weak measurements has been developed quite generally for all observables, but for didactic reasons, we shall stick to weak measurements of position, which are relevant for experiments to measure trajectories. And since the measured trajectories are Bohmian, we shall remain for the moment in the realm of Bohmian mechanics.

We return to (7.32), but model the apparatus now by a pointer wave function $\Phi = \Phi(y)$, while $\varphi(x)$ stands for the initial wave function of the particle. For the sake of simplicity we consider only a one-dimensional situation and describe the pointer by a single degree of freedom. That is, $Y \in \mathbb{R}$ is the position of the pointer and $X(t) \in \mathbb{R}$ is the position of the particle at time t. For concreteness, we may think of

$$\Phi(y) \sim e^{-y^2/4\sigma^2} , \tag{8.1}$$

---

[1] W. Heisenberg, *Der Teil und das Ganze: Gespräche im Umkreis der Atomphysik.* Piper, 1996, S. 80. Translation by the authors.

© Springer Nature Switzerland AG 2020

D. Dürr, D. Lazarovici, *Understanding Quantum Mechanics*,

https://doi.org/10.1007/978-3-030-40068-2_8

so that

$$\int y\,|\Phi(y)|^2\,\mathrm{d}y = 0\,. \tag{8.2}$$

The pointer wave function is thus centred around the ready state 0 and has "width" σ. We now consider an unsharp position measurement procedure with probabilities given by (7.24). Concretely (but not very realistically), we can think of the so-called von Neumann measurement with the temporal evolution given by

$$U(\Delta t) = \exp\left(-\mathrm{i}\frac{\Delta t}{\hbar}\hat{X}\hat{P}_y\right)\,,$$

where $\hat{X}$ denotes the position operator of the particle and $\hat{P}_y$ denotes the momentum operator of the pointer. Δt is the duration of the interaction between particle and apparatus. The initial wave function $\varphi(x)\Phi(y)$ then evolves in time according to

$$U(\Delta t)\varphi(x)\Phi(y) = \exp\left(-\mathrm{i}\frac{\Delta t}{\hbar}\hat{X}\hat{P}_y\right)\varphi(x)\Phi(y)$$

$$= \exp\left(-\Delta tx\frac{\partial}{\partial y}\right)\varphi(x)\Phi(y) = \varphi(x)\Phi(y - \Delta tx)\,.$$

The last equality comes from the Taylor expansion

$$\Phi(y - \Delta tx) = \Phi(y) - \Delta tx\frac{\partial}{\partial y}\Phi(y) + \frac{1}{2}\left(-\Delta tx\frac{\partial}{\partial y}\right)^2\Phi(y) + \cdots\,.$$

We set $t = -1$ as the initial time and $t = 0$ for the end of the interaction, i.e., $\Delta t = 1$, and hence, at the end of the measurement process, we obtain

$$\varphi(x)\Phi(y) \xrightarrow{U(1)} \varphi(x)\Phi(y - x)\,. \tag{8.3}$$

The pointer position Y is distributed according to Born's statistical law

$$\rho^Y(y) = \int |\varphi(x)|^2|\Phi(y - x)|^2\mathrm{d}x\,, \tag{8.4}$$

and with this probability, for pointer position Y, the wave function collapses to

$$\varphi_{0+}(x) = \varphi_Y(x) := \varphi(x)\Phi(Y - x)\,. \tag{8.5}$$

Note here that we usually think of a position measurement as being made by apparatus which displays the position with high precision. This would be the case for very small σ in (8.1). Since $\Phi(Y-x) \approx 0$ for $|Y-x| \gg \sigma$, for small σ, the wave function (8.5) would be an approximate eigenfunction of $\hat{X}$, rather sharply peaked around $x = Y$.

But in a weak measurement the pointer wave function is very broad, i.e., σ is much greater than the width of φ, which we can think of as being close to zero in comparison with σ. In that sense, $\Phi(Y - x)$ does not change significantly on the scale on which the x-values vary in (8.5). Hence, to a good approximation, we can set

$$\varphi_Y(x) \approx \Phi(Y)\varphi(x) . \tag{8.6}$$

Normalizing, we get the effective wave function

$$\varphi_{0+}(x) \approx \varphi(x) , \tag{8.7}$$

where the error is of the order of $1/\sigma$. The moral is that the wave function of the particle is only barely changed by the measurement process.

We can look upon the "weakness" of the measurement in another way. In typical measurements, we assume that the pointer wave functions corresponding to different measurement results have macroscopically disjoint supports, which yields decoherence for the system wave functions. Because of the spread of the pointer wave function considered here, even for significantly different positions $Y_1 \neq Y_2$, the wave functions $\Phi(Y_1 - x)$ and $\Phi(Y_2 - x)$ will have a considerable overlap (in x), and the system wave function will not "collapse".

What is there to gain? In a single measurement, nothing. Some pointer position comes out which is very weakly correlated with the actual position of the particle. But under a great many repetitions of the measurement experiment, we do at least get the empirical mean of the pointer positions by virtue of (8.4) and a simple computation using (8.2):

$$\mathbb{E}(Y) \equiv \int y\rho^Y(y)\mathrm{d}y = \int x\rho^X(x)\mathrm{d}x \equiv \mathbb{E}(X) ,$$

where $\rho^X(x) = |\varphi(x)|^2$. That is already more than nothing, although not much more. But now comes the trick which turns weak measurements into a powerful tool.

First observe that, if a "strong measurement" of the position is performed immediately after each weak measurement, and if we now consider probabilities only for sub-ensembles, let's say one in which the strong measurement result is $\tilde{X}$, then we may argue in the following way. Considering the conditional probability density for Y given $X = \tilde{X}$, i.e.,

$$\rho^Y(y\,|X = \tilde{X}) = \frac{\rho^{X,Y}(\tilde{X}, y)}{\rho^X(\tilde{X})} = \frac{|\varphi(\tilde{X})|^2|\Phi(y - \tilde{X})|^2}{|\varphi(\tilde{X})|^2} = |\Phi(y - \tilde{X})|^2, \qquad (8.8)$$

and taking into account (8.1) and (8.2), we get the theoretical expected value

$$\mathbb{E}(Y|X = \tilde{X}) \equiv \int y\rho^Y(y\,|X = \tilde{X})\,dy = \tilde{X}, \qquad (8.9)$$

for the empirical mean of this sub-ensemble. In other words, if we consider all those events in which the particle has been detected at the position $\tilde{X}$ (which in Bohmian mechanics means all the events in which the particle *was in fact* at the position $\tilde{X}$), then that position is reliably registered by a sequence of weak measurements. That is still not so very exciting. But now let some small time τ elapse between the weak and the strong measurement. Then the empirical mean of the weak measurements is the approximate position, say before the duration τ, at time $t = 0$, and after τ the position is known "exactly". We consider a sub-ensemble as before, but we now condition under $X(\tau) = \tilde{X}$ and consider the velocity formula

$$\lim_{\tau \to 0} \frac{1}{\tau}\mathbb{E}(\tilde{X} - Y|X(\tau) = \tilde{X}) = \lim_{\tau \to 0} \frac{1}{\tau}\Big[\tilde{X} - \mathbb{E}(Y|X(\tau) = \tilde{X})\Big], \qquad (8.10)$$

where the expected value is taken at the initial time 0. We can compute that! We use the Bohmian velocity v^φ [see, e.g., (4.6)] and, for $X(0) = X$, we get approximately

$$X(\tau) \approx X + v^{\varphi_\tau}(X(\tau))\tau. \qquad (8.11)$$

Now when τ tends to 0 and the weak measurement is made at time 0, we can use (8.7) to get

$$v^{\varphi_\tau} \approx v^{\varphi_{0+}} \approx v^\varphi, \qquad (8.12)$$

where φ is the initial wave function of the particle before the measurement, and hence,[2]

$$X(\tau) \approx X + v^\varphi(X(\tau))\tau. \qquad (8.13)$$

---

[2]Note that the replacement [see (8.7)] comes with an error $1/\sigma$ and the reader might wonder whether a more careful argument on the order of τ is needed. Indeed it is, but for ease of presentation we leave it as is and refer to D. Dürr, S. Goldstein, and N. Zanghì, On the Weak Measurement of Velocity in Bohmian Mechanics. In: *Quantum Physics Without Quantum Philosophy*, Springer, 2013, Sect. 7.3, for a more detailed analysis.

The event $X(\tau) = \tilde{X}$ can be identified with the event $X = \tilde{X} - v^\varphi(\tilde{X})\tau$ in this approximation, so using (8.9), we get

$$\mathbb{E}(Y|X(\tau) = \tilde{X}) \approx \mathbb{E}(Y|X = \tilde{X} - v^\varphi(\tilde{X})\tau) = \tilde{X} - v^\varphi(\tilde{X})\tau . \qquad (8.14)$$

Plugging this into (8.10), we find that the sequence of measurements does in fact yield information about the Bohmian velocity, namely,

$$\lim_{\tau \to 0} \frac{1}{\tau}\left[x - \mathbb{E}(Y|X(\tau) = x)\right] = v^\varphi(x) , \qquad (8.15)$$

where x is now arbitrary.

Weak measurements of the velocity have been carried out experimentally.[3] In this way, we can reconstruct trajectories which are of particular interest for the double slit experiment. This should be contrasted with the still often expressed view that the interference pattern on the screen behind the double slit proves that trajectories cannot exist.

Look now at the two pictures of trajectories in Figs. 8.1 and 8.2, where the first is taken from the experiment mentioned above (see the reference in footnote 3). To understand the reconstruction of the curves, we should think of the space behind the double slit as being foliated into planes, in each of which the velocities are measured in the prescribed way at sufficiently many points x. After many repetitions, we obtain velocity vectors in each plane filling out the space behind the slits. We can then construct the velocity vector field and plot its integral curves.

The second figure shows the theoretically computed trajectories in Bohmian mechanics, where the wave function parts in the slits are modelled by Gaussian wave packets. The strangely curving trajectories shortly after the slits result from the interference of the wave packets at the upper and lower slits. The straight lines later on emerge from the transition of the wave function into plane wave packets, as discussed in Sect. 1.3.

Note that, in both pictures, trajectories which go through the upper slit never cross the symmetry axis. They stay above it. Likewise for the trajectories which go through the lower slit. This is inevitable for Bohmian trajectories, as they are determined by a vector field and are hence solutions of a first order differential equation [see (4.6)]. Trajectories inherit the mirror symmetry of the experiment: if one trajectory crossed the symmetry axis, there would be a mirrored trajectory, and at their point of crossing there would be two velocity vectors. This would violate the property of a velocity vector field according to which there is a unique vector at every point in space. In short, different trajectories cannot cross.

---

[3]S. Kocsis, B. Braverman, S. Ravets, M.J. Stevens, R.P. Mirin, L.K. Shalm, and A.M. Steinberg, Observing the average trajectories of single photons in a two-slit interferometer. Science **332**, 1170–1173 (2011).

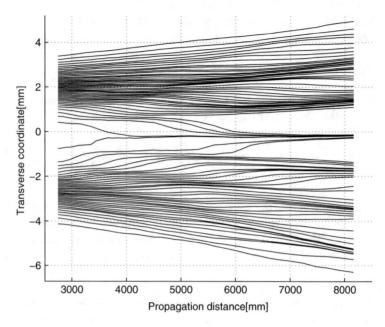

Fig. 8.1 Reconstruction of the photon trajectories in a double-slit experiment by weak measurement of the velocities. The *vertical axis*, which is also the alignment of the double slit, is measured in millimeters. The *horizontal axis* represents the distance from the double slit, also in millimeters. From Kocsis et al.

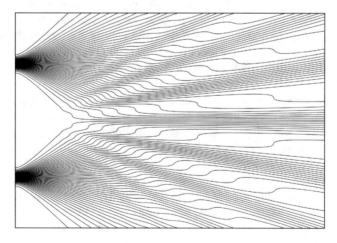

Fig. 8.2 Numerical simulation of Bohmian trajectories in a double-slit experiment. Graphic by Leopold Kellers

8.1   On the (Im)-Possibility of Measuring the Velocity

We have made two seemingly contradictory statements. In Remark 7.3, we showed that the velocity of Bohmian particles cannot be measured, and in this chapter we claim the opposite. How can this be? Underlying this is the ambiguity in the notion of measurement, which we already alluded to in Remark 7.3. The negative statement concerned measurement experiments like the one in (7.24). The weak measurement with its subsequent strong measurement and the conditioning on a sub-ensemble which selects suitable measurement results *a posteriori* is of a different kind. If we extend the notion of measurement accordingly, then it is indeed justified to say that, from the viewpoint of Bohmian mechanics, we can measure the Bohmian velocities—and hence also the particle trajectories.

Here it is worth recalling Einstein's wise warning that it is theory that decides what is measurable. However, the data obtained in the weak position measurement are at the end of the day due to macroscopic pointer positions. And since other quantum theories (like GRW theory or Many Worlds theory if the latter obeys Born's statistical law) describe the statistics of pointer positions correctly, they will also predict the results of the weak measurement experiments correctly, i.e., they will predict that the measurement results will produce the Bohmian velocity vector field. But in the framework of such theories, one must of course deny that the trajectories reconstructed from the results of the measurement experiments (as presented in Fig. 8.1) correspond to trajectories of real particles.

Having said that, we note that (8.15) can also be derived from the operator formalism without appeal to Bohmian mechanics.[4] In Dirac notation, it takes the form

$$\mathbb{E}\big(Y|X(\tau)=x\big) = \operatorname{Re}\frac{\langle x(\tau)|\hat{X}|\varphi\rangle}{\langle x(\tau)\mid\varphi\rangle} = \operatorname{Re}\frac{\langle x|U(\tau)\hat{X}|\varphi\rangle}{\langle x|U(\tau)|\varphi\rangle}. \tag{8.16}$$

Here $U(t)$ is the free evolution

$$U(t) = e^{-\frac{i}{\hbar}tH}, \quad \text{where} \quad \hat{H} = \frac{\hat{p}^2}{2m} = -\frac{\hbar^2}{2m}\Delta_x.$$

From that, we readily compute (good exercise)

$$\lim_{\tau\to 0}\frac{1}{\tau}\Big[x - \mathbb{E}\big(Y|X(\tau)=x\big)\Big] = \operatorname{Re}\frac{\langle x|\frac{i}{\hbar}[\hat{H},\hat{X}]|\varphi\rangle}{\langle x\mid\varphi\rangle} = \operatorname{Im}\frac{\hbar}{m}\frac{\nabla_x\varphi(x)}{\varphi(x)} = v^\varphi(x). \tag{8.17}$$

---

[4]H.M. Wiseman: Grounding Bohmian mechanics in weak values and Bayesianism. New Journal of Physics **9**, 165 (2007). The experiment referred to in footnote 3 is actually based on Wiseman's suggestion.

In this operational view, it is agreed that the left-hand side of (8.17) is a "velocity", but we cannot ask "velocity of what?"

Bohmian mechanics yields the most natural explanation of the described experiment: the measured velocities are actually the particle velocities according to Bohmian mechanics, and the reconstructed trajectories correspond likewise to the trajectories of a Bohmian particle ensemble. However, the weak measurements of the Bohmian velocities cannot falsify other quantum theories, because other theories (GRW and Many Worlds) make the same predictions for the pointer statistics as in (8.17).

8.2   Surrealistic Trajectories?

We consider the Mach–Zehnder interferometer shown in Fig. 8.3. Photons are sent through a beam splitter which transmits half of the beam and deflects the other half orthogonally downwards. In both interferometer arms, the photons eventually reach a mirror which deflects them in the direction of one detector. At the end of the experiment we register which of the detectors D_1 and D_2 has clicked.

The Schrödinger evolution of the wave function[5] is easily followed in this case. The initial wave function splits into two parts at the beam splitter—let us call them ϕ_1 and ϕ_2. These propagate along the upper and lower arms of the interferometer, respectively, and entangle with the detector wave functions. At the point where they cross one another, the wave functions superpose for a short time and a second beam splitter could be placed there to erase one of the two phases through interference—in the following, however, we consider only the situation without the second beam splitter.

The different quantum theories yield different descriptions of what is going on in this experiment. According to the Many Worlds theory, there are only the two wave packets moving in the interferometer. If we call those wave packets "particles", the particle propagates through both arms simultaneously. At the end of the experiment, both detectors click, but within any given world, the experimenter there sees only one of the detectors click. So we end up with two worlds, and in one world D_1 clicks and in the other D_2.

According to GRW theory, the photon wave function collapses with almost complete certainty when the wave packets entangle with the detector wave function, and as a matter of fact, only one of the detectors clicks. What happens inside the interferometer depends now on the choice of the ontology of the GRW theory. If the ontology is mass density, then one half of the density moves through the upper arm and one half through the lower arm, and at the moment of collapse the mass density contracts to the place of the detector which clicks. This is a radically nonlocal effect: the mass density which was moving towards the other detector spontaneously

---

[5]Unfortunately, there is no agreement about what the wave function of a photon is. We ignore here this "little" detail for the sake of argument.

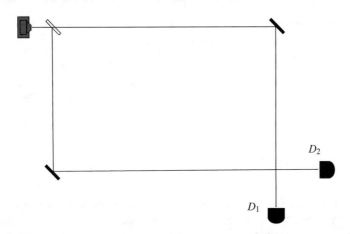

Fig. 8.3 Mach–Zehnder interferometer. The photon or particle source (*upper left*) sends one photon (one wave) through a beam splitter. The separated wave parts will cross by the action of mirrors (*lower left* and *upper right*) and the wave parts will be detected by the detectors (*lower right*)

delocalises, so that shortly after the click the whole mass density is at the place of the detector which clicked. In GRWf theory, the arms of the interferometer would with high probability be empty—i.e., strictly speaking there is nothing propagating between source and detector. The subsequent collapse produces a flash event inside the detector, making it click. This would be described in words by saying that a "photon arrived in that detector".

According to Bohmian mechanics, a point particle moves on a well defined trajectory through the interferometer, either in the upper arm or in the lower arm, and it arrives at one detector which then clicks. Once again, only one of the detectors actually clicks. In Bohmian mechanics (and only in this theory), it makes sense to ask which trajectory the particle takes through the interferometer to arrive at detector D_1 or D_2.

Intuitively (or better, naively), one would think that particles which arrive at D_1 went through the upper arm of the interferometer and that particles which arrive at D_2 went through the lower arm. But Bohmian particles behave differently. Just recall that different Bohmian trajectories cannot cross each other (see Fig. 8.2 and its explanation). The only possibility is therefore that the particles move on the trajectories depicted in Fig. 8.5.

That is what the theory says and that is what is happening if the theory is correct. There is nothing to wonder or ponder about. It is simply the case that the trajectories depicted in Fig. 8.4 are based on a classical (Newtonian) intuition which is inappropriate when it comes to the quantum regime. To get a better appreciation of this, let us elaborate on the Bohmian evolution. After the particle has passed the beam splitter, it travels with either the upper or the lower wave packet (ϕ_1 or ϕ_2). At the crossing point where the two wave packets are superposed, the particle

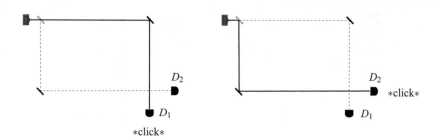

Fig. 8.4 "Naive" (classical) trajectories in the Mach–Zehnder interferometer

changes its guiding wave packet. This is because, due to the symmetry of the set up, there must be trajectories that end up at each detector, and as in the double slit situation, the two possible trajectories can't cross. Hence, after the crossing point, particles which first moved with the upper wave packet ϕ_1 will be guided by ϕ_2 and arrive at D_2. Vice versa for the particles which were initially guided by ϕ_2. The relative frequencies are those predicted by Born's rule and they are corroborated in experiments. So far, so good.

It is, of course, possible to mistrust all theoretical descriptions and instead try to make an operational reconstruction of the photon trajectory by placing a measurement device in the arms of the interferometer. A normal strong measurement of position is out of the question, because that would result in a collapse of the wave function and would change the entire set up. So it is better to try with weak measurements as discussed above. Thus we do weak position measurements in both arms, followed by strong measurements at the detectors which finally absorb the particle. A single photon will not leave a recognizable trace in the weak measurement. But after repeating the experiment a great many times, we consider the statistics of the weak measurement outcomes, once for the sub-ensemble of particles which have arrived at detector D_1 and once for the sub-ensemble which have arrived at D_2. We then ask in which arm of the interferometer the particles which arrived at D_1 left a weak trace, and likewise for D_2. This is called a *which-way detection*.

Doing the experiment, it is found that, in the sub-ensemble where detector D_1 clicks, there is a weak trace in the upper arm, while in the other sub-ensemble, the opposite happens. In other words, the "which-way detection" agrees with the "naive" trajectories of Fig. 8.4 and disagrees with the Bohmian trajectories of Fig. 8.5. More complicated setups can be invented and the weakly measured paths sometimes agree with the Bohmian ones, sometimes with the "naive" ones, and sometimes with neither.[6] In fact, the right correlation between the weak measurements and the final detection is found by following the propagation of the wave packets which couple to the wave function of the clicking detector.

---

[6]See, for instance, L. Vaidmann, Past of a quantum particle. Physical Review A **87**, 052104 (2013).

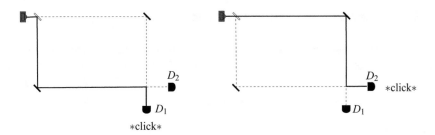

Fig. 8.5 Bohmian trajectories in the Mach–Zehnder interferometer

The take-home message here is that the notion of "which-way detection" is misleading. Giving it this name makes it sound as though there is a path taken by a particle, but for that, the theory must contain particles in the first place, otherwise what would we actually be talking about? For example, in the Copenhagen interpretation of quantum mechanics, there are no variables representing particle positions or trajectories, so it doesn't make sense to ask which way the particle went.

Should anyone insist on an operational notion of "paths", drawing on intuitions that are not grounded in any precise theory, they can of course create conflicts with theories that do actually describe particle trajectories. An example can be found in the paper *Surrealistic Bohm trajectories*,[7] where the disagreement between weakly registered paths and the theoretical Bohmian paths led the authors to call the Bohmian trajectories "surrealistic", while it would have been more appropriate to call the which-way detections surrealistic, since they do not in general detect the way followed by anything. Bohmian mechanics correctly predicts the phenomena of the weak measurements, where the value displayed by the weak detector is caused by the wave function, which is also part of Bohmian mechanics. But in contrast to the situation described in the previous section, the theory no longer predicts that a series of weak measurements will accurately register the position of the particles as they are passing through the interferometer.

Succinctly put, real particles need not behave like dogs, leaving a trace at every tree they pass. In particular, it should be realized that the intuitions which nourish naive which-way detection operationalism are based on the principle of locality, which happens to be falsified by quantum phenomena. In the end, the take-home message is once again Einstein's insight: it is theory that decides what is measurable and how it can be measured.

---

[7]B.-G. Englert, M.O. Scully, G. Süssmann, and H. Walther, Surrealistic Bohm trajectories. Zeitschrift für Naturforschung **47a**, 1175 (1992).

8.3   Wheeler's Delayed-Choice Experiment

The interferometer setup can be used to depict quite a lot of interesting and less interesting quantum phenomena. We wish to end this chapter with one often debated but nevertheless quite nonsensical setup, in which a second beam splitter is set up on the crossing point of the Mach–Zehnder interferometer and adjusted in such a way that interference allows only one beam to remain, let's say in the direction of detector D_2.

In the outdated way of speaking of "wave–particle dualism", it is sometimes said that, if no second beam splitter is inserted, then the particle takes only one of the possible paths through the interferometer (because at the end, only one detector clicks), but if the second beam splitter is inserted, then the particle must go through both arms of the interferometer, because it interferes destructively "with itself" at the second beam splitter. This confused description can now be capped by the following "clever" observation. If we act fast enough (and the interferometer is big enough), we can wait until the particle has passed the first beam splitter and only then decide whether or not to insert the second beam splitter. This setup is called a "delayed choice" experiment, following John Archibald Wheeler (1911–2008). What does this delayed choice achieve? It allegedly causes retroaction, influencing the past of the particle, i.e., it changes the past! Because when the second beam splitter is inserted and destructive interference takes place, the particle must have gone through both arms. But it was already on its way through one of the arms after passing through the first beam splitter and before the second beam splitter was put in place.

It is absurd that this is still considered a mystery, even after decades, during which hundreds of publications have been written. But this is what happens when no clear and precise formulation of quantum mechanics is adopted.

Hidden Variables

9

A final moral concerns terminology. Why did such serious people take so seriously axioms which now seem so arbitrary? I suspect that they were misled by the pernicious misuse of the word "measurement" in contemporary theory. This word very strongly suggests the ascertaining of some preexisting property of some thing, any instrument involved playing a purely passive role. Quantum experiments are just not like that, as we learned especially from Bohr. The results have to be regarded as the joint product of "system" and "apparatus", the complete experimental setup. But the misuse of the word "measurement" makes it easy to forget this and then to expect that the "results of measurements" should obey some simple logic in which the apparatus is not mentioned. The resulting difficulties soon show that any such logic is not ordinary logic. It is my impression that the whole vast subject of "Quantum Logic" has arisen in this way from the misuse of a word. I am convinced that the word "measurement" has now been so abused that the field would be significantly advanced by banning its use altogether, in favour for example of the word "experiment".

John S. Bell[1]

Every course on quantum mechanics will at some point engage with the notion of "hidden variables". The terminology is connected with the infamous *no hidden variables theorems* of von Neumann, Gleason, Kochen and Specker, and Bell, which assert that quantum mechanics does not allow for hidden variables.

Before we go into that, we first wish to say something about the terminology itself. It originates in von Neumann's famous book *Mathematische Grundlagen der Quantenmechanik*.[2] As discussed in the previous chapters, orthodox quantum mechanics focus on the wave function and the operator observables. We have already understood that the role of the operator observables is to act as bookkeepers for the statistics of measurement experiments. That is what von Neumann developed in his book attributing importance, in particular, to projector-valued measures (PVMs). But having elements in our most fundamental description of Nature that

---

[1]J.S. Bell, *Speakable and Unspeakable in Quantum Mechanics*, 2nd edn, Cambridge University Press, Cambridge, 2004, p. 166.

[2]Springer, 1932.

© Springer Nature Switzerland AG 2020
D. Dürr, D. Lazarovici, *Understanding Quantum Mechanics*,
https://doi.org/10.1007/978-3-030-40068-2_9

are merely bookkeepers of statistics seems perhaps a bit defeatist. The idea of "measuring observables" in an experiment subsequently became part of quantum folklore.

For example, it is common practice to say that a Stern–Gerlach magnet measures the spin observable—in short, that we measure the spin of an electron. This way of speaking can seduce practitioners into a naive realism about operators in which it is believed that, when we measure an observable, we do in fact measure something that is already "there", i.e., a property that the system possesses independently of the measurement experiment performed on it. In other words, the idea would be that the measurement simply reveals the value of a variable which is already present. Just as when we measure our collar size, its size is already there, it's just that we don't know its exact value until it is revealed by the measurement.

Hence, when the observable $\hat{A}$ is measured and the measurement value is α_k (an eigenvalue of the observable), i.e., when the apparatus displays the value α_k, we consider that a function (in general, a coarse-graining function)

$$Z_A : \Omega \mapsto \{\text{eigenvalues of } \hat{A}\}$$

exists on a state space Ω such that $Z_A(\omega) = \alpha_k$ was the preexisting value before the measurement (for example, the factual size of someone's neck), which is determined by the state space variable ω. Coarse-graining functions are referred to, as we know, as random variables.

Example Consider an electron in the spin state $\alpha|\uparrow_z\rangle + \beta|\downarrow_z\rangle$, which is sent through a Stern–Gerlach magnet oriented in the z-direction. The measurement yields z-spin $+1/2$ with probability $|\alpha|^2$ and z-spin $-1/2$ with probability $|\beta|^2$. The wave function "collapses" to the corresponding eigenvector, let's say $|\uparrow_z\rangle$. Then we say that the particle "has z-spin $+1/2$". Furthermore, this can be checked by a subsequent "z-spin measurement". Speaking that way, we may be tricked into thinking that the particle had z-spin $+1/2$ all along, even before the first measurement, and that we merely learned this by doing the measurement. In this way of thinking, the wave function $\alpha|\uparrow_z\rangle + \beta|\downarrow_z\rangle$ would then merely represent our ignorance: the electron *has* z-spin $+1/2$ with probability $|\alpha|^2$ and z-spin $-1/2$ with probability $|\beta|^2$, and it's just that we don't know which of the two. Mathematically, this would be expressed by attributing a random variable Z_{σ_z} whose values $Z_{\sigma_z}(\omega) \in \{\pm 1/2\}$ correspond to the possible measurement values. $\omega \in \Omega$ would represent a "hidden" state variable which determines the measured spin value. "Randomness" would then come from not knowing the value of ω, and hence $Z_{\sigma_z}(\omega)$, before the measurement, while the distribution $\rho(\omega)$ had to be such that, in a typical measurement sequence, the quantum mechanical probabilities would arise, i.e., $\mathbb{P}(Z_{\sigma_z} = +1/2) = \int \mathbb{1}_{\{Z_{\sigma_z}=+1/2\}} \rho(\omega) \, d\omega = |\alpha|^2$.

Such a detailed description in terms of variables $Z(\omega)$, $\omega \in \Omega$ [like phase space coordinates $\omega = (q, p)$, for example, in classical physics] is not part of orthodox or Copenhagen quantum mechanics, so von Neumann referred to them as "hidden" variables. Whether we call Z or ω the hidden variable does not really matter because Z and ω come as a package, and there is no harm calling them both hidden variables: they are variables that may still remain to be found in the description of Nature and would be responsible for the outcomes of measurements.

Von Neumann asked himself: Is it possible for such hidden variables $Z(\omega)$ to exist or would their existence be in contradiction with quantum mechanics? For Werner Heisenberg and Niels Bohr, the answer was clearly: existence is a no-go. Von Neumann, a mathematician, wanted to secure the negative answer and designed and proved a theorem—the first of its kind. It was a *no hidden variables* theorem. After that came many more. We shall discuss von Neumann's theorem and one more recent version of a *no hidden variables* theorem at the end of the chapter. Another version will be proven in Chap. 10 on Bell's inequality, although there, the question is a completely different one. Moreover, the latter is the only theorem of its kind that is truly relevant and we shall explain why.

We need only start by succinctly and very generally formulating the prototype of a *no hidden variables theorem*:

Theorem 9.1 (Prototype of a No Hidden Variables Theorem) *There does not exist a "good map"*

$$\hat{A} \mapsto Z_A \tag{9.1}$$

from the self-adjoint operators on a Hilbert space $\mathscr{H}$ to random variables on an ordinary probability space Ω, where $Z_A = Z_A(\omega)$ is to be taken as a possible measurement value, i.e., as an eigenvalue of $\hat{A}$.

We admit that the expression "good map" is not overly precise, but it is nevertheless adequate here, because there are various requirements that could be imposed on the mapping and each set of assumptions yields another no hidden variables theorem. Later, we shall give some examples of what might be considered as "good". On the other hand, all concepts agree that the quantum mechanical probabilities for measurement values of an observable, say $\hat{A}$, should equal the probability distribution of Z_A. However, that is not so important right now. To see its irrelevance, it suffices to recall the role of the observable $\hat{A} = \sum \alpha_k P_k$ as bookkeeper of the statistics of measured values α_k. Let $\mathscr{E}$ be a measurement experiment in which the apparatus displays the values α_k and in which the statistics of the values is encoded in the spectral resolution P_k, as discussed in Chap. 7. In short, we have the map [compare with (7.22)]

$$\mathscr{E} \mapsto \hat{A},$$

which is defined by the values and their statistics. But that implies that the map is not injective: many experiments with completely different setups will yield the same values and their statistics, which are encoded in the same $\hat{A}$. To make a caricature of this, if we take a computer which prints the values α_k with relative frequencies which agree with the probabilities computed from the spectral resolution and the wave function as in (7.14), then that is an experiment which "measures" $\hat{A}$.

From this it becomes intuitively clear that the encoding of the statistics by an observable is too abstract and too coarse to describe the physical situation in a measurement experiment, let alone stand for a fundamental physical property with which the observable may be associated. In other words, the non-existence of the function (9.1) originates from the fact that there are in general many random variables—depending on the details of the measurement experiment at hand—to which the observable would have to be mapped. So, loosely speaking, what we have is rather like a one-to-many map, except that it is not a map, since the functional prescription is ill-defined. Actually, not much more needs to be said as the status of relevance of the theorem is now clear and a formal proof becomes unnecessary. But we shall do it anyway.

In particular, the case of Bohmian mechanics shows that it is possible to measure a preexisting value, for example, in a position measurement, which measures the actual position of the Bohmian particle. Besides position, Bohmian mechanics contains no other random variables which exist independently of the measurement experiment—as shown by the "measurement" of the momentum observable (see Chap. 4). Moreover, not every measurement experiment whose statistics are described by the position operator measures the actual position of the Bohmian particle. We discussed that in Chap. 8 when we looked at "surrealistic trajectories".

9.1   Joint Measurements of Observables

As it happens, the formal proofs of the variants of Theorem 9.1 do not refer to the non-uniqueness. It is something that only surfaces indirectly, as we shall explain now. This will also allow us to point out some further peculiarities in the handling of operator observables. To do this, we need to talk about joint probabilities, which arise in sequential "measurements of observables". We use the notation of Chap. 7.

We "measure" one observable after the other. That is, we have two pieces of measurement apparatus Φ_k and Ψ_l and we "measure" first with the Φ_k-apparatus (measurement A) and then with the Ψ_l-apparatus (measurement B). To get the probabilities for the displays, we need only recall that it is the Schrödinger evolution of the total system, two pieces of apparatus and the measured system, that we have to consider. Thus (7.3) is relevant once again, except that the wave function at the end is now

$$\sum_{k,l} \varphi_{k,l} \Phi_k \Psi_l \, , \tag{9.2}$$

and for simplicity we have kept the new effective wave functions $\varphi_{k,l} := P_l^B P_k^A \varphi$ non-normalized, which is why the $c_{k,l}$ don't show up. The computation (7.4) can be repeated to show that the pair of values α_k, β_l has probability

$$\mathbb{P}^\varphi(\alpha_k, \beta_l) = \| P_l^B P_k^A \varphi \|^2 , \tag{9.3}$$

with orthogonal projectors as in Chap. 7.

We start with an experiment $\mathscr{E}$, associated with the spectral resolution $P_k, k \in I$, where I is an index set. We can have many different pieces of apparatus with different sets of values on their displays. For example, we can have $\mathscr{A} = \{\alpha_k\}$ and $\mathscr{B} = \{\beta_k\}$. This means that we can simultaneously perform two "measurements" associated with two bookkeepers $\hat{A} = \sum_{k \in I} \alpha_k P_k$ and $\hat{B} = \sum_{k \in I} \beta_k P_k$. Then, with (9.3), we have

$$\begin{aligned}
\mathbb{P}^\varphi(\alpha_k, \beta_l) &= \| P_l P_k \varphi \|^2 \\
&= \langle P_l P_k \varphi, P_l P_k \varphi \rangle \\
&= \langle \varphi, P_l P_k \varphi \rangle ,
\end{aligned} \tag{9.4}$$

where the last equality is due to the properties of orthogonal projectors, i.e., $P^2 = P$, and self-adjointness $P^+ = P$, and also commutativity $[P_k, P_l] =: P_k P_l - P_l P_k = 0$ (which is automatically satisfied for a family of orthogonal projectors). We do the calculation

$$\begin{aligned}
\langle P_l P_k \varphi, P_l P_k \varphi \rangle &= \langle \varphi, (P_l P_k)^+ P_l P_k \varphi \rangle \\
&= \langle \varphi, P_k^+ P_l^+ P_l P_k \varphi \rangle \\
&= \langle \varphi, P_k P_l P_l P_k \varphi \rangle \\
&= \langle \varphi, P_k P_l P_k \varphi \rangle \\
&= \langle \varphi, P_l P_k P_k \varphi \rangle = \langle \varphi, P_l P_k \varphi \rangle .
\end{aligned}$$

Of course, the value set $\mathscr{B}$ can also be coarser than $\mathscr{A}$. It could, for example, be a subset of the latter. In any case, the values β are then "degenerate eigenvalues", i.e., we have a coarse-grained subset $(P_l')_{l \in I'}$ of the spectral family $(P_k)_{k \in I}$, with projections onto subspaces of higher dimensions. Here $I' \subset I$ is the index set for $\mathscr{B}$ and $(P_l')_{l \in I'}$ is now the corresponding spectral family, i.e., $\hat{B} = \sum_{l \in I'} \beta_l P_l'$. More generally, the α values and β values could both be degenerate with different index sets I and I', but as long as the corresponding families $(P_k)_{k \in I}$ and $(P_l')_{l \in I'}$ commute, there are essentially no changes.[3] All we need for simultaneous

---

[3]For instance, if $\hat{A} = \hat{H}$—the "energy operator"—and $\hat{B} = \hat{L}$—the "angular momentum operator"—we have, in general, different energy levels for every value of $\hat{L}$ (and vice versa). A joint measurement then leads to channelling into common eigenspaces.

"measurability" is commutativity of the families, i.e.,

$$[P_k, P_l'] = 0, \quad \text{for all } k, l,$$

This is easily seen. The family $P_{k,l} := P_k P_l'$ is itself a spectral resolution and moreover $[\hat{A}, \hat{B}] = 0$, i.e., the observables commute. Conversely, it is a standard result of linear algebra that commuting self-adjoint operators $\hat{A}$, $\hat{B}$ have a common spectral resolution, that is, they can be diagonalized in the same basis. Furthermore,

$$\sum_{k \in I} P_{k,l} = P_l', \quad \sum_l P_{k,l} = P_k, \quad \sum_{k \in I, l \in I'} P_{k,l} = 1. \tag{9.5}$$

Plugging this into (9.4) and using linearity of the scalar product, we obtain

$$\sum_{\alpha_k \in \mathscr{A}} \mathbb{P}^\varphi(\alpha_k, \beta_l) = \mathbb{P}^\varphi(\beta_l), \tag{9.6}$$

$$\sum_{\beta_l \in \mathscr{B}} \mathbb{P}^\varphi(\alpha_k, \beta_l) = \mathbb{P}^\varphi(\alpha_k), \tag{9.7}$$

$$\sum_{\alpha_k \in \mathscr{A}, \beta_l \in \mathscr{B}} \mathbb{P}^\varphi(\alpha_k, \beta_l) = 1. \tag{9.8}$$

This says that we have a consistent (i.e., just a normal) family of joint probabilities. This consideration extends without further ado to N commuting observables. The moral which every student of quantum mechanics learns is thus that only commuting observables are simultaneously "measurable".

To make this a little less abstract, we may think this way. The spectral resolution of the observable corresponds in an experiment to the splitting of the wave function into channels on the configuration space of the system and apparatus. Only if the observables $\hat{A}$ and $\hat{B}$ commute and hence have a common spectral resolution does there exist a consistent channeling corresponding to the possible measurement values of $\hat{A}$ and $\hat{B}$ which is independent of the order in which the measurements were made.

Let us consider now the general situation, where the bookkeepers are non-commuting observables $\hat{A}$, $\hat{B}$. We first "measure" $\hat{A} = \sum_k \alpha_k P_k^A$ and then $\hat{B} = \sum_l \beta_l P_l^B$. Once again, we have the probabilities (9.3), but now

$$\mathbb{P}(\alpha_k, \beta_l) = \| P_l^B P_k^A \varphi \|^2 \neq \langle \varphi, P_l^B P_k^A \varphi \rangle, \tag{9.9}$$

because the families P_k^A and P_l^B fail to commute. For this reason, (9.6) does not hold. Equations (9.7) and (9.8) do hold, but the order of the measurements is important.

Remark 9.1 (Another Example of a POVM) The sequence of measurements just discussed cannot be associated with a spectral resolution, so it cannot be associated with an observable. But it is a measurement nevertheless, at least in the sense that some values are pointed out at the end of the day. Thus if we were to insist that only self-adjoint operator observables are "measurable", we would be missing out on some possibly important situations. Anyway, we already know what to say. This measurement experiment can still be associated with a POVM, so nothing really new is going on, except that we have another very natural example of a POVM.

As a consequence, we find that no consistent family of joint probabilities exists: the marginal distributions do not result as in (9.6) from summation over the values we wish to ignore, because (9.6) just does not hold. Is that exciting? We sum over all possible values of the first measurement, that is we ignore the outcomes, and we do not get the probabilities for the single "measurement" of $\hat{B}$. So we must accept (and that should by now be easy enough) that by simply ignoring the first measurement experiment we cannot undo what the first measurement experiment did to the wave function when "measuring" $\hat{A}$.

The formula (9.9) is sometimes referred to as the Wigner formula,[4] but it was found independently by many others.[5] The take-home message is that joint probabilities exist for commuting observables, but there is no consistent family of joint probabilities for non-commuting observables. This is more or less the only content of the infamous no hidden variables theorems.

Take for example three observables $\hat{A}, \hat{B}, \hat{C}$, where $\hat{A}$ commutes with $\hat{B}$ and $\hat{C}$, but $\hat{B}$ does not commute with $\hat{C}$. Then there exist joint probabilities for $\hat{A}$ and $\hat{B}$ as well as for $\hat{A}$ and $\hat{C}$, while no joint probabilities exist for $\hat{B}$ and $\hat{C}$. On the other hand, random variables Z_A, Z_B, Z_C always have joint probabilities. If we now require the map from observables to random variables to be "good" (and that needs to be made precise, but we may already guess one way to do that), we can imagine that something must go wrong.

Our first intuition about what is going on in Theorem 9.1 was that it had something to do with non-uniqueness, i.e., the fact that many different experiments yield the same statistics. Where does this non-uniqueness hide in the present discussion? It is just the possibility of "measuring" the observable $\hat{A}$ in different ways which are incompatible with each other: for example, once simultaneously with $\hat{B}$ and once simultaneously with $\hat{C}$.

In the next section, we shall prove two such no hidden variables theorems, von Neumann's and another due to Kochen and Specker, where we present the much simplified proof by Mermin. We note for later reference that, in the proof of Bell's inequality (Chap. 10), we cannot assign random variables to the six spin observables which enter the proof. This is indeed another no hidden variables result, but one

---

[4]Eugene Wigner (1902–1995).

[5]The formula also served as the starting point for the so-called decoherent (or consistent) history approaches to quantum physics.

which uses a very particular quantum state—a singlet wave function—while the former results do not hinge on any particular choice of wave function.

9.2   Two Assertions About Hidden Variables

9.2.1   Von Neumann's Theorem

The theorem requires the map (9.1) to be linear and concludes that hidden variables don't exist. If $\hat{C} = \hat{A} + \hat{B}$, then we know that the quantum mechanical mean values satisfy $\langle \hat{C} \rangle = \langle \hat{A} \rangle + \langle \hat{B} \rangle$. Von Neumann, succumbing to standard quantum mechanical usage, called the hidden variables ω "dispersion free states". Their existence would then imply that, instead of the linearity of the quantum mechanical mean values, we should have $Z_{A+B} = Z_A + Z_B$ for the random variables. The condition for a "good" map (9.1) is thus

$$\hat{C} = \hat{A} + \hat{B} \implies Z_C(\omega) = Z_A(\omega) + Z_B(\omega) \text{ for almost all } \omega \in \Omega . \qquad (9.10)$$

If the Hilbert space $\mathcal{H}$ has at least dimension two, then there can be no such map on the set of self-adjoint operators on $\mathcal{H}$.

Proof The values of $Z_{A,B,C}$ must be eigenvalues of $\hat{A}$, $\hat{B}$, $\hat{C}$. But the eigenvalues of $\hat{A} + \hat{B}$ are in general not the sum of the eigenvalues of $\hat{A}$ and $\hat{B}$. A simple example of two non-commuting 2×2 matrices is sufficient to see that—a simple exercise for the reader.

The proof is correct, so the theorem is true. And now what? What does this have to do with physics, and in particular with quantum physics? Why is the requirement of linearity reasonable? Think for example of the position observable $\hat{X}$ and the momentum observable $\hat{P}$. What would a "measurement" of $\hat{X} + \hat{P}$ actually mean? Because of its stark physical irrelevance, the von Neumann theorem is sometimes described as naive or silly in more recent literature.

9.2.2   The Kochen–Specker Theorem

This theorem requires something less naive for the map $\hat{A} \mapsto Z_A$ than the linearity imposed by von Neumann, viz.,

Definition 9.1 *The map (9.1) can be described as "good", or in the usual jargon, non-contextual, if the following is satisfied: whenever quantum mechanical joint probabilities for a set of self-adjoint operators $(\hat{A}_1, \ldots, \hat{A}_m)$ exist, i.e., whenever they form a commuting family, they agree with the joint probabilities of the corresponding random variables $(Z_{A_1}, \ldots, Z_{A_m})$.*

As a consequence, all algebraic identities which hold between the commuting observables must also hold for the random variables. For example, if $\hat{A}$, $\hat{B}$, and $\hat{C}$ is a commuting family and $\hat{C} = \hat{A}\hat{B}$, then also $Z_C = Z_A Z_B$, because the joint probability of Z_A, Z_B, and Z_C must be 0 for the set of values $\{(\alpha, \beta, \gamma) \in \mathbb{R}^3 | \gamma \neq \alpha\beta\}$.

As in von Neumann's requirement, we would like the relations which hold for operators to hold also for the hypothetical hidden random variables, the difference being that the requirement is now restricted to families of observables which can be jointly measured. Once again, we can find an example which shows that such a map from observables to random variables does not exist.

Proof We consider a four-dimensional Hilbert space[6] and take as observables the Pauli matrices for two independent spin-1/2 particles σ_a^1 and σ_b^2. The algebra of observables is in this case characterized as follows:

- For all directions a and for $k = 1, 2$, we have $(\sigma_a^k)^2 = \mathbf{1}$, i.e., the eigenvalues are ± 1.
- For all directions a, b, we have $\sigma_a^1 \sigma_b^2 - \sigma_a^2 \sigma_b^1 = 0$.
- If a, b are orthogonal directions, then $\sigma_a^k \sigma_b^k + \sigma_a^k \sigma_b^k = 0$ for $k = 1, 2$.
- For $a = x$, $b = y$, we have $\sigma_x^k \sigma_y^k = i\sigma_z^k$ for $k = 1, 2$.

Consider now the following scheme of observables:

$$\sigma_x^1 \quad \sigma_x^2 \quad \sigma_x^1 \sigma_x^2$$

$$\sigma_y^2 \quad \sigma_y^1 \quad \sigma_y^1 \sigma_y^2$$

$$\sigma_x^1 \sigma_y^2 \quad \sigma_y^1 \sigma_x^2 \quad \sigma_z^1 \sigma_z^2$$

Using the algebraic relations, it is straightforward to check that:

a) The observables along each column and each row commute.
b) The product along each row is 1.
c) The product along the first column and along the second column is 1, while that along the third column is -1.

As a consequence of Definition (9.1), all relations between commuting observables must be reproduced by the values of the map (9.1). This means that, for each ω, we must be able to attribute the values $+1$ or -1 to the nine observables in such a way

---

[6]The four-dimensional Hilbert space is the tensor product of two two-dimensional ones, one per particle. Then to be mathematically rigorous, σ_a^1 should be thought of as, e.g., $\sigma_a^1 \otimes \mathbf{1}$, and σ_b^2 would then be $\mathbf{1} \otimes \sigma_b^2$, where $\mathbf{1}$ is the two-by-two unit matrix.

that b) and c) are jointly satisfied. But that is impossible, because multiplying first all the rows, the product of the nine values would have to be $+1$, while multiplying all the columns, the product of the nine values would have to be -1. This proves Theorem 9.1 for Kochen–Specker goodness.

9.3   Contextuality

The Kochen–Specker theorem is often interpreted as proving the impossibility of introducing *non-contextual* hidden variables into quantum mechanics without coming into conflict with the empirical predictions. What is the meaning of this term? The question of hidden variables is often associated with the measurement problem of quantum mechanics, which we discussed in detail in Chap. 2, and consequently with the question of the completion of orthodox quantum theory. To describe these completing variables—which we called *beables* or *the ontology*—as *hidden* is rather odd, however. In order to solve the measurement problem, these variables must not be hidden at all. They must rather be the variables that are responsible for the readily *perceivable* difference between a pointer pointing to the left and a pointer pointing to the right.

Nevertheless, since the early quantum theory did not contain such variables, there was a strong tendency to describe these possibly missing pieces as hidden. Even David Bohm, in his famous paper of 1952 which led eventually to Bohmian mechanics,[7] used the terminology of hidden variables for particle positions— unfortunately, one must say, because of the misunderstanding that the suggested theory presented a refutation of von Neumann's theorem. In reality, Bohmian mechanics only shows that the theorem is physically irrelevant.

Notwithstanding, the no-go theorems à la von Neumann and Kochen–Specker are still discussed with the intention of proving that a completion of orthodox quantum mechanics is impossible or would be accompanied by all sorts of absurd consequences. The theorems are supposed to stipulate, as it were, that the quantum mechanical measurement formalism tells us everything there is to say about Nature and that any search for a completion of the theory would be in vain. But isn't it rather absurd to insist that quantum mechanics is only about measurable quantities, while at the same time we must add that there is nothing objective at all that is actually being measured? In any case, such a dogmatic reading of the mathematical results is simply not justified. Bohmian mechanics and the GRW theory prove that it is perfectly possible and meaningful to supplement the wave function by ontological quantities which describe the real state of the system, i.e., the actual matter configuration. But note that these *beables* do not need to have any other properties besides positions. To ask for more than that endangers naive realism about

---

[7]D. Bohm, A suggested interpretation of the quantum theory in terms of 'hidden' variables I and II. Physical Review **85**, 166, 180 (1952).

observables! One should always bear in mind Bohr's dictum that it is the experiment which creates the measured values—position measurements being the exception.

Think, for example, of the Bohmian description of spin measurements. It would be wrong to say that the measured spin corresponds to an intrinsic property of the particle which it has in addition to its position. It is the Stern–Gerlach magnet we set up which makes the wave function split into the corresponding spinor components that subsequently separate spatially, in such a way that the particle follows one of the two wave components. The situation is similar in GRW and the Many Worlds theory. After the splitting, it is the interaction with the measurement apparatus that ensures that the spin components decohere and entangle with macroscopically distinguishable pointer positions. In collapse theory, the entanglement also ensures that the common wave function of the particle and apparatus collapses and thus "selects" one of the possible measurement results.

So what is the purpose of the discussion on contextuality? Instead of throwing up another philosophical molehill and calling non-existent properties "contextual", the best thing is to agree with Niels Bohr: observables do not generally correspond to properties with pre-determined values which are revealed in a measurement. If, however, we insist on associating observables with physical properties of a system, then these properties necessarily depend on the context in which we intend to measure it. This is the ambiguity that comes to bear in the proof of the no-go theorem. As above, $\hat{A}$ can be measured with $\hat{B}$ and also with $\hat{C}$, but not with $\hat{B}$ and $\hat{C}$ together. The hypothetical property at hand which is supposed to be associated with the observable $\hat{A}$ would then be a different one when $\hat{A}$ is measured with $\hat{B}$ than when it is measured with $\hat{C}$. Should that shock us? Not really, because the measurement of $\hat{A}$ with $\hat{B}$ generally requires a completely different experiment than the measurement of $\hat{A}$ with $\hat{C}$. The observable is always just a shorthand in which the complete measurement experiment, including the measurement setup, no matter how complex, simply disappears. And all the mysteries, from contextuality to quantum logic, only arise if we try to think of this shorthand as being something fundamental, pretending that there is no physics left in the measuring process itself.

The final irony of this whole contextuality story is that contextuality has a negative connotation for everybody. That is, basically everybody agrees that contextuality is elevated nonsense. So why is it still being spread around? Mainly to suppress any attempt to complete quantum theory by, e.g., Bohmian mechanics— no matter what. The argument being that such completions contain contextual properties and therefore are in some way sinful.

Nonlocality

10

> I cannot believe in it [quantum mechanics], because the theory is incompatible with the principle that physics is supposed to represent a reality in space and time, without spooky action at a distance.
>
> Albert Einstein, Letter to Max Born on March 3, 1947. Translation by the authors.

What has been said so far is all well and good, but the implications of the quantum phenomena and the measurement problem of orthodox quantum mechanics are far from conclusive. We appear to be left with many options: determinism (Bohm, Everett) or indeterminism (GRW), many worlds or a single one, particles or flashes or the wave function alone—all these possibilities could, in principle, fit physical reality as we know it. And now we turn to something new, something fundamental, in fact, an eternal truth: the nonlocality of Nature. We will make the concept more precise as we go along, but in a nutshell, nonlocality means that the fundamental laws of Nature must involve some sort of action at a distance, i.e., distant events sometimes influencing each other instantaneously.[1]

Newtonian mechanics is an action-at-a-distance theory: it involves absolute simultaneity, which allows for instantaneous interactions. The nonlocality of Newtonian mechanics is, however, rather mild, because the strength of interactions decays quickly with increasing distance.

Bohmian mechanics, on the other hand, is strikingly nonlocal: the dynamics are formulated on configuration space, where all particles are guided together and simultaneously by a common wave function. In a two-particle system with coordinates $\mathbf{X}_1(t)$ and $\mathbf{X}_2(t)$, we have

$$\dot{\mathbf{X}}_1(t) = -\frac{\hbar}{m_1} \nabla_{\mathbf{x}} \mathrm{Im} \ln \psi\big(\mathbf{x}, \mathbf{X}_2(t)\big)\big|_{\mathbf{x}=\mathbf{X}_1(t)},$$

---

[1] "Instantaneous" provides the right intuitive understanding, but we should note that it is no longer well defined when we come to relativistic physics. "Superluminal" would be a more precise way of speaking, but it actually fails to express just how radical this new feature is (see footnote 10).

© Springer Nature Switzerland AG 2020
D. Dürr, D. Lazarovici, *Understanding Quantum Mechanics*,
https://doi.org/10.1007/978-3-030-40068-2_10

whence the velocity of X_1 at time t depends on the position of X_2 *at the same time*, and vice versa, as long as the wave function $\psi(x, y)$ is entangled, i.e., as long as it doesn't factorize into a product.

The GRW theory is also manifestly nonlocal. Acting on one part of an entangled system can affect the collapse rate and thus lead to an instantaneous change in the density of matter (flashes) in an arbitrarily distant location. (For Everett, the issue is more subtle, as discussed in Sect. 6.3.)

The critical question is now the following: Is this nonlocality necessary or is it merely a peculiarity of these particular theories, i.e., would it be possible in principle to provide a local description of Nature if we had some other theory? More succinctly, is Nature itself local or nonlocal? A bold question, indeed, but John Stewart Bell had the audacity to address it, and propose an experimental test that could establish the nonlocality of Nature once and for all! It is important to appreciate the depth and scope of this achievement: an experiment that tells us how Nature has to be described, no matter what. A truth that stands above any theory. Quite rightly, the nonlocality of Nature has been called "the most profound discovery in science".[2] When all is said and done, it is *the* key innovation of quantum physics, and it is absolutely impossible to understand quantum mechanics without understanding nonlocality.

10.1   The EPR Argument

Einstein was probably the first to realize that the entanglement of the wave function in quantum mechanics leads to a nonlocal description of Nature. The violation of the locality principle was, in fact, at the center of his rejection of quantum mechanics— not the apparent violation of determinism, as is often claimed with reference to his famous quote that "God doesn't play dice".[3]

Since for Einstein, the principle of locality was non-negotiable, he concluded that quantum theory could not be a complete description of Nature. This is the point of the famous article by Einstein, Podolsky, and Rosen (EPR), which shows that the assumption of locality implies the incompleteness of quantum mechanics.[4] We shall consider Bohm's version of the EPR experiment (also called the EPRB experiment), which Bell discussed as well and which is essentially what is realized in actual experiments. We recall Sect. 1.7 in which we discussed the propagation of a spinor wave function in a Stern–Gerlach magnet. For the EPRB experiment, a special pair

---

[2] H.P. Stapp, Bell's theorem and world process. Nuovo Cimento B **29**, issue 2, 270–276 1975.

[3] A more complete quote can be found in Einstein's letter to Cornelius Lanczos: "It seems hard to sneak a look at God's cards. But that he plays dice and uses 'telepathic' methods (as the present quantum theory requires of him) is something that I cannot believe for a moment." The " telepathic methods" are a reference to the action at a distance that Einstein identified in quantum mechanics.

[4] A. Einstein, B. Podolsky, and N. Rosen, Can quantum-mechanical description of physical reality be considered complete? Physical Review **47**, 777 (1935).

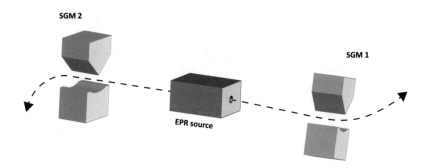

Fig. 10.1 The EPRB experiment. Two particles in the singlet state (10.1) fly off in opposite directions through rotatable Stern–Gerlach devices SGM 1 and SGM 2. The particles are thereby deflected towards either the flat or the pointy pole piece. Rotation of the magnets happens while the particles are in flight, and so quickly that no light signal could "communicate" the orientation of the magnet to the particle at the other side of the experiment

of spin-1/2 particles, called an EPR pair, is prepared in the so-called singlet state. This is an entangled state with total spin zero. The spin part of the antisymmetric wave function has the form

$$\psi_s = \frac{1}{\sqrt{2}} \left(|\uparrow\rangle_1 |\downarrow\rangle_2 - |\downarrow\rangle_1 |\uparrow\rangle_2 \right), \tag{10.1}$$

where $|\uparrow\rangle_i$ and $|\downarrow\rangle_i$ are the spin eigenstates of the i th particle, and we can neglect the direction because the state is rotationally symmetric.

The crucial point as that this two-particle state has the following property: if we measure the spin of one of the particles in any direction **a**, we get "**a**-spin UP" or "**a**-spin DOWN" each with probability 1/2, but if we measure the spin of both particles in the same direction, we always obtain opposite results, i.e., if the **a**-spin of particle 1 is UP, then the **a**-spin of particle 2 is always DOWN, and vice versa. We have already talked about the meaning of the expression "having spin UP or DOWN". It refers only to the phenomenon that the particle is deflected either towards the sharp pole of the magnet or the flat one as it passes through the inhomogeneous magnetic field inside the Stern–Gerlach apparatus before it hits a detector screen (see Fig. 10.1).[5]

Put simply, if the two particles move apart through Stern–Gerlach magnets with the same orientation, they are always deflected in opposite directions. This is a simple physical fact that doesn't seem too suspicious. But now we can move the two Stern–Gerlach magnets SGM 1 and SGM 2 as far apart as we like (at least in principle, since decoherence always threatens to spoil the experiment), so far

---

[5]When Stern first proposed these experiments, he suggested using a magnet which was oriented from the floor to the ceiling of the lab. This may be responsible for the terminology "up" and "down".

that no influence propagating at most with the speed of light could act between the measurement on particle 1 and the measurement on particle 2 (the two measurement events are "spacelike separated", to use the relativistic terminology). Still, if we observe that particle 1 has "**a**-spin UP" (let's say), we can immediately infer that particle 2 has "**a**-spin DOWN", i.e., that a subsequent measurement of the **a**-spin on particle 2 must yield the outcome "spin DOWN".

The innocent reader who has not yet had much opportunity to study quantum mechanics and still has plenty of common sense won't find this particularly suspicious, either. The particles "just have opposite spins", she will say, "so, if I observe that particle 1 has **a**-spin UP, I can conclude that particle 2 will have **a**-spin DOWN." Indeed, the whole phenomenon would be utterly trivial if we could suppose that the particles had their spin values all along, and that these values were then simply revealed by the measurements.

The problem emphasized by EPR arises, however, because quantum mechanics insists that the spin values of the particles are *not* determined in advance, but come about through the process of measurement (and the collapse of the wave function). This would mean that the **a**-spin of particle 2 is *physically* determined by the measurement on particle 1 (supposing that the measurement on particle 1 occurs first), and that measuring "**a**-spin UP" for particle 1 affects the physical state in such a way that a corresponding measurement on particle 2 must yield "**a**-spin DOWN", even though any non-superluminal influence between the two measurements is excluded.

In other words, the EPR dilemma is the following. The first possibility is that particle 2 had "**a**-spin DOWN" all along, independently of the measurement performed on particle 1, in which case the quantum mechanical description is incomplete and there must exist additional variables determining the outcomes of the spin measurements. The second possibility is that the **a**-spin of particle 2 *does* depend on the measurement process carried out on particle 1. In that case, the measurement of "**a**-spin UP" for particle 1 is what causes a subsequent measurement on particle 2 to yield "**a**-spin DOWN", whence there must be some form of action at a distance, i.e., nonlocality.

It's very important to think this EPR argument through, since there is so much unnecessary controversy and confusion about it. In particular, it's important to note that (leaving a many-worlds scenario aside), there is really no third option available. The EPR correlations require either "instantaneous" influences over arbitrary distances, or additional variables absent in the quantum mechanical description. That's what Einstein, Podolsky, and Rosen had already proven before Bell came on the scene.

10.2 Bell Inequality

So let's suppose that Nature is local. Then, the spins of the two EPR particles must have been determined prior to the measurements. And since we can choose the orientations of the two Stern–Gerlach magnets at the very last moment—so that these choices cannot have any local influence on the opposite side of the experiment—this must be true for the spins in *any* direction.

Formally, this means that there exists a family of random variables ("hidden variables" in the sense of Chap. 9)

$$Z_{\mathbf{a}_1}^{(1)}, Z_{\mathbf{a}_2}^{(2)} \in \{-1, 1\}, \quad \text{such that} \quad \mathbf{a}_1 = \mathbf{a}_2 \implies Z_{\mathbf{a}_1}^{(1)} = -Z_{\mathbf{a}_2}^{(2)} \tag{10.2}$$

whose values determine the results of the spin measurements on particles 1 and 2 in arbitrary directions $\mathbf{a}_1$ and $\mathbf{a}_2$, and which are correlated in such a way as to produce the right empirical frequencies.

The mathematical term "random variable" is somewhat misleading. Note that we (or better, Mark Kac) already complained about that notion. A random variable is a function, usually a coarse-graining function. The functions $Z_{\mathbf{a}_1}^{(1)}, Z_{\mathbf{a}_2}^{(2)} \in \{-1, 1\}$ may represent an intrinsically probabilistic law but could also depend deterministically on other physical quantities. It really doesn't matter what the theory that is supposed to explain the spin correlations actually looks like. It can be deterministic or indeterministic, nice or ugly, simple or hopelessly complex. The only assumption is that it reproduces the statistics of the EPR experiment in a *local* manner, and the variables $Z_{\mathbf{a}_1}^{(1)}, Z_{\mathbf{a}_2}^{(2)}$ merely describe the predictions of this hypothetical theory for the outcomes of the spin measurements. Notably, under the assumption of locality, $Z_{\mathbf{a}_1}^{(1)}$ cannot depend on $\mathbf{a}_2$ and vice versa, since these choices can also be made at "spacelike separation", while the particles are already in flight. Therefore, $Z_{\mathbf{a}_1}^{(1)}$ and $Z_{\mathbf{a}_2}^{(2)}$ are also referred to as *local* hidden variables.

We now choose three arbitrary directions $\mathbf{a}, \mathbf{b}, \mathbf{c}$ and consider the values

$$(Z_{\mathbf{a}}^{(1)}, Z_{\mathbf{b}}^{(1)}, Z_{\mathbf{c}}^{(1)}) = (-Z_{\mathbf{a}}^{(2)}, -Z_{\mathbf{b}}^{(2)}, -Z_{\mathbf{c}}^{(2)}). \tag{10.3}$$

We consider the probabilities of the anti-coincidences $Z_{\mathbf{a}}^{(1)} = -Z_{\mathbf{b}}^{(2)}$, $Z_{\mathbf{b}}^{(1)} = -Z_{\mathbf{c}}^{(2)}$, and $Z_{\mathbf{a}}^{(1)} = -Z_{\mathbf{c}}^{(2)}$, and sum them up. Thus, we obtain

$$\mathbb{P}(Z_{\mathbf{a}}^{(1)} = -Z_{\mathbf{b}}^{(2)}) + \mathbb{P}(Z_{\mathbf{b}}^{(1)} = -Z_{\mathbf{c}}^{(2)}) + \mathbb{P}(Z_{\mathbf{c}}^{(1)} = -Z_{\mathbf{a}}^{(2)})$$

$$\overset{(10.3)}{=} \mathbb{P}(Z_{\mathbf{a}}^{(1)} = Z_{\mathbf{b}}^{(1)}) + \mathbb{P}(Z_{\mathbf{b}}^{(1)} = Z_{\mathbf{c}}^{(1)}) + \mathbb{P}(Z_{\mathbf{c}}^{(1)} = Z_{\mathbf{a}}^{(1)})$$

$$\geq \mathbb{P}(Z_{\mathbf{a}}^{(1)} = Z_{\mathbf{b}}^{(1)} \text{ or } Z_{\mathbf{b}}^{(1)} = Z_{\mathbf{c}}^{(1)} \text{ or } Z_{\mathbf{c}}^{(1)} = Z_{\mathbf{a}}^{(1)})$$

$$= \mathbb{P}(\text{"certain event"})$$

$$= 1,$$

since $Z_{\mathbf{a},\mathbf{b},\mathbf{c}}^{(i)}$ can only take the values $+1$ and -1, and this means that one of the cases $Z_{\mathbf{a}}^{(1)} = Z_{\mathbf{b}}^{(1)}$ or $Z_{\mathbf{b}}^{(1)} = Z_{\mathbf{c}}^{(1)}$ or $Z_{\mathbf{c}}^{(1)} = Z_{\mathbf{a}}^{(1)}$ must always hold true. This is one version of the famous *Bell inequality*:

$$\mathbb{P}(Z_{\mathbf{a}}^{(1)} = -Z_{\mathbf{b}}^{(2)}) + \mathbb{P}(Z_{\mathbf{b}}^{(1)} = -Z_{\mathbf{c}}^{(2)}) + \mathbb{P}(Z_{\mathbf{c}}^{(1)} = -Z_{\mathbf{a}}^{(2)}) \geq 1. \tag{10.4}$$

It is obviously a rather trivial consequence of the existence of the random variables, i.e., a direct implication of the locality assumption.

The only thing left to do is to carry out the experiment. This seems almost like a pointless enterprise. The statement looks so trivial that it can only be confirmed. And yet, all the precise quantum theories that we have encountered so far were nonlocal. And if the principle of locality is put into question, then so is the existence of the random variables $Z_{\mathbf{a}}^{(1)}$ and $Z_{\mathbf{a}}^{(2)}$, and the relative frequencies could add up to something less than 1. Indeed, how the experiment pays off! For the sum of the relative frequencies—corresponding to the left-hand-side of (10.4)—we obtain a value significantly smaller than 1 (we will discuss recent experimental results below). As a consequence, the Bell inequality is violated and experiment excludes *any* conceivable theory which tries to explain the spin (anti-)correlations without nonlocal influences. In other words, experiment rules in favor of nonlocality!

At the same time, the experiments carried out so far all confirm the quantum mechanical predictions, which are the following. As mentioned before, the spin-singlet wave function is

$$\psi_s = \frac{1}{\sqrt{2}}(|\uparrow\rangle_1|\downarrow\rangle_2 - |\downarrow\rangle_1|\uparrow\rangle_2),$$

ignoring the spatial part $\psi(x_1, x_2) = \psi(x_2, x_1)$ which is simply multiplied by ψ_s. Now we compute the expected value of the coincidences, viz.,

$$E_{\mathbf{a},\mathbf{b}} = \frac{1}{4}\langle\psi_s|\mathbf{a}\cdot\boldsymbol{\sigma}^{(1)}\otimes\mathbf{b}\cdot\boldsymbol{\sigma}^{(2)}|\psi_s\rangle.$$

This expression is bilinear in $\mathbf{a}, \mathbf{b}$ and rotationally invariant, hence a multiple of $\mathbf{a}\cdot\mathbf{b}$. That is, $E_{\mathbf{a},\mathbf{b}} := \mu\mathbf{a}\cdot\mathbf{b}$, for some μ, and we can read off the proportionality factor μ from the simple case $\mathbf{a} = \mathbf{b}$, where $E_{\mathbf{a},\mathbf{b}} = -1/4$. Hence

$$E_{\mathbf{a},\mathbf{b}} = -\frac{1}{4}\mathbf{a}\cdot\mathbf{b}. \tag{10.5}$$

On the other hand, writing $P_{\mathbf{a},\mathbf{b}}$ for the probability of the anti-coincidences (the probability that the $\mathbf{a}$-spin of particle 1 is opposite to the $\mathbf{b}$-spin of particle 2, i.e., the probability that the product of their spin values is $-1/4$), we have

$$E_{\mathbf{a},\mathbf{b}} = -\frac{1}{4}P_{\mathbf{a},\mathbf{b}} + \frac{1}{4}(1 - P_{\mathbf{a},\mathbf{b}}) = -\frac{1}{2}P_{\mathbf{a},\mathbf{b}} + \frac{1}{4},$$

and therefore,

$$P_{\mathbf{a},\mathbf{b}} = \frac{1}{2} + \frac{1}{2}\mathbf{a}\cdot\mathbf{b}. \tag{10.6}$$

Choosing **a**, **b**, **c** with intermediate angles of 120°, we find

$$P_{\mathbf{a},\mathbf{b}} = \frac{1}{2} - \frac{1}{4} = \frac{1}{4}, \quad P_{\mathbf{a},\mathbf{c}} = \frac{1}{4}, \quad P_{\mathbf{b},\mathbf{c}} = \frac{1}{4}.$$

Hence, for the left-hand-side of (10.4), we obtain the number 3/4 and thus a clear violation of the Bell inequality.

10.3 Implications and Misunderstandings

Einstein was of course right in his conviction that the quantum theory of his time was incomplete. The measurement problem discussed in Chap. 2 makes this abundantly clear. The only thing was that it had nothing to do with nonlocality. Bohm, GRW, and Everett offer different ways to "complete" standard quantum mechanics but none of them would have been (or were) to Einstein's liking as they violate the principle of locality.

John Bell thus wanted to know if it was possible, at least in principle, to provide a local description of quantum phenomena. The answer he found was negative. The fact that the predictions of quantum mechanics violate the Bell inequality means that these predictions cannot be reproduced by *any* local theory. And the fact that experiments confirm these predictions—in particular, the violation of the Bell inequality—means that *no* local theory can correctly describe our world. The nonlocality of Nature has thus been established once and for all.

Notably, as we will discuss in detail below, this nonlocality does not imply the possibility of faster-than-light signaling—which would strongly clash with Einstein's theory of relativity. Nor does it imply a return to Newtonian physics, with forces acting instantaneously throughout all of space. It does, however, imply that certain statistical correlations observed in Nature cannot be explained without admitting that distant events can directly influence each other—even if these events occur at such long spatial and short temporal separations that any non-superluminal interactions between them are excluded.

Bohmian mechanics explains nonlocal correlations by entanglement of the wave function and a nonlocal law of motion for particles. GRW explains them by entanglement of the wave function and a nonlocal (stochastic) law for the localization of matter. The Many Worlds theory explains them (or, some would say, explains them away) by the entanglement of the wave function and a branching history of the universe. Other descriptions are conceivable, but Einstein's principle of locality has been shattered once and for all.

Bell's argument is so clear and precise that we may well wonder how it could have given rise to so many debates and misunderstandings. That a great number of physicists fail to understand "the most profound discovery" of their discipline is tragic, but it is a fact that we still have to contend with today. The most common mistake is to read Bell's theorem as just another "no-hidden-variables" result along the lines of those discussed in Chap. 9. It is then often said that the violation of Bell's theorem implies that we have to give up either locality or "realism". (The latter is a grossly inadequate term and the reader should ask how such a philosophical concept could be of any relevance here.)

As a matter of fact, if we ask whether we should be "naive realists" about the spin observables, meaning that the particles could have "hidden spin values" $Z_{\mathbf{a}_1}^{(1)}$ and $Z_{\mathbf{a}_1}^{(2)}$ that are revealed in experiments, the answer provided by Bell's theorem is clearly negative: random variables $Z_{\mathbf{a}_1}^{(1)}$ and $Z_{\mathbf{a}_2}^{(2)}$ reproducing the quantum correlations (10.5) are mathematically impossible. However, to regard this as the main message of Bell's theorem is to miss the point entirely, because it forgets about the EPR argument which was the very basis of Bell's investigation. By the EPR argument, the existence of the random variables $Z_{\mathbf{a}_1}^{(1)}$ and $Z_{\mathbf{a}_2}^{(2)}$ was *inferred* from the assumption of locality. The impossibility of such variables thus implies that the assumption of locality must be violated. We are left with only one of the alternatives of the EPR dilemma, namely nonlocality.

Let us emphasize this again. If we measure "**a**-spin UP" for particle 1 and can infer that a corresponding measurement on particle 2 must yield "**a**-spin DOWN", we have indeed not merely learned about a pre-existing property that particle 2 had independently of our interaction with particle 1. But this means precisely that the measurement on particle 1 must be, in some sense, responsible for the fact that a subsequent measurement on particle 2 will yield "**a**-spin DOWN", no matter how far away it occurs. What's more, the situation is symmetric between particles 1 and 2, and we can already see the difficulties arising in the relativistic context when it no longer makes sense to ask which of the two measurements occurred first.

Bell himself repeatedly protested against the widespread misunderstanding of his argument. In footnote 10 of his famous article *Bertlmann's socks and the nature of reality*, he writes[6]:

> My own first paper on this subject [Physics **1**, 195 (1965)] starts with a summary of the EPR argument from locality to deterministic hidden variables. But the commentators have almost universally reported that it begins with deterministic hidden variables. (p. 157)

And in the article itself:

> It is important to note that to the limited degree to which determinism plays a role in the EPR argument, it is not assumed but inferred. What is held sacred is the principle of 'local causality'—or 'no action at a distance'. [...] It is remarkably difficult to get this point across, that determinism is not a presupposition of the analysis. (p. 143)

Here, "determinism" refers to the existence of local hidden variables determining the outcomes of the spin measurements, i.e., what many authors mean by "realism" (if they mean anything precise at all).

---

[6]Reprinted in: J.S. Bell, Speakable and Unspeakable in Quantum Mechanics, 2nd edn, Cambridge University Press, 2004, Chap. 16.

Schematically, the logical structure of Bell's argument is as follows:

EPR   locality $\Longrightarrow$ local hidden variables

Bell   local hidden variables $\Longrightarrow$ Bell inequality

Experiment   ¬Bell inequality $\Longrightarrow$ ¬local hidden variables $\Longrightarrow$ ¬locality

Starting with the EPR argument, the only physical assumption underlying the derivation of Bell's inequality is locality. The violation of Bell's inequality observed in various experiments thus means that the principle of locality is falsified.

10.4   CHSH Inequality and the Generalized Bell Theorem

At the beginning of this chapter, we promised an insight that stands "above any theory". Our discussion so far, important as it may be, does not *quite* live up to this promise, since the derivation of the Bell inequality (10.4) assumed perfect spin anti-correlations (10.3). Quantum mechanics predicts these perfect anti-correlations for an EPR pair in the singlet-state, but quantum mechanics could be wrong, putting the inference from the violation of Bell's inequality to nonlocality into question. Actual experiments are consistent with the predictions of quantum mechanics but will never be able to show that (10.3) is true with absolute certainty.

For this reason, we will now discuss a more general version of Bell's theorem that does not require the assumption of perfect anti-correlations. This is based on the so-called CHSH inequality, due to Clauser, Horne, Shimony, and Holt. It is the violation of this CHSH inequality that is actually reported in the relevant experiments. In the course of this discussion, we will also provide more rigorous definitions of the relevant concepts—in particular "locality".

The starting point of Bell's analysis is the prediction, and subsequent experimental observation, of statistical correlations between spacelike separated events A and B. "Spacelike separated" means that no signal, propagating at most with the speed of light, could be sent from one event to the other. The existence of statistical correlations means that the joint probability of $\mathbb{P}(A, B)$ does not factorize:

$$\mathbb{P}(A, B) \neq \mathbb{P}(A) \cdot \mathbb{P}(B) \,. \tag{10.7}$$

Alternatively, we may consider the *conditional probability* $\mathbb{P}(A \mid B) :=$ $\mathbb{P}(A, B)/\mathbb{P}(B)$. Equation (10.7) is then equivalent to

$$\mathbb{P}(A \mid B) \neq \mathbb{P}(A) \tag{10.8}$$

and

$$\mathbb{P}(B \mid A) \neq \mathbb{P}(B) \,. \tag{10.9}$$

In other words, conditioning on B increases or decreases the probability for the occurrence of A and vice versa—the two events are statistically dependent.

In the EPRB experiment, we have the following situation. Let $(A \mid a) \in \{\pm1\}$ denote the outcome of the spin measurement on particle 1 in the direction a ("spin UP" or "spin DOWN"), and $(B \mid b) \in \{\pm1\}$ the result of the spin measurement on particle 2 in the direction b.[7] In the spin singlet state, we have: $\mathbb{P}(A = +1 \mid a) = \mathbb{P}(B = +1 \mid b) = 1/2$, but with (10.6)

$$\mathbb{P}(A = +1, B = +1 \mid a, b) = \frac{1}{4}(1 - \mathbf{a} \cdot \mathbf{b}). \tag{10.10}$$

Unless $\mathbf{a}$ and $\mathbf{b}$ are orthogonal directions, i.e., $\mathbf{a} \cdot \mathbf{b} \neq 0$, we thus have

$$\mathbb{P}(A = +1, B = +1 \mid a, b) \neq \mathbb{P}(A = +1 \mid a) \cdot \mathbb{P}(B = +1 \mid b), \tag{10.11}$$

so that the outcomes of the distant spin measurements are (anti-)correlated.

In themselves, such correlations between distant events are very common and do not indicate any particular causal relation. In particular, we wouldn't generally be surprised to find correlations between jointly prepared systems no matter how far they have subsequently been separated. In a *local* theory, however, a complete description of the physical state in the past will contain all the relevant information for the prediction of A and B, whence conditioning also on the occurrence of B would become redundant for prediction of A, and vice versa.

We can consider rather banal cases: the number of car accidents that occur on a given day in New York is statistically correlated with the number of car accidents that occur on the same day in Seattle. Evidently, this does not mean that a car crash in New York can "cause" an accident on the West Coast, or vice versa. Instead, weather conditions are often comparable in the two places, people all over the country tend to consume more alcohol on Saturdays than during the week, etc. Once we condition on all factors that could serve as a *local explanation* or *common cause* for an increased number of car crashes, we will find that the accident statistics for New York become independent of simultaneous events in Seattle, and vice versa.

Remark 10.1 (On Statistical Independence) In Chap. 3 on chance in physics, we repeatedly emphasized that the statistical independence of relevant coarse-grainings ("random variables") is a very tricky issue. For statistical independence to hold, the pre-images of the coarse-graining functions have to mix and intertwine perfectly (recall Fig. 3.1.) The Rademacher functions r_k, $k = 1, 2, \ldots$ on the interval $[0, 1)$, equipped with the Lebesgue measure were our prototypes of coarse-graining

---

[7]The parameters a, b, etc., still represent the orientations $\mathbf{a}$, $\mathbf{b}$, etc., of the spin measurements, as in the previous section. However, we shall drop the vectorial notion from now on, since it is usual to think of the relevant parameters as angles in the plane of rotation of the Stern–Gerlach magnets (orthogonal to the flight trajectory of the particles).

variables. We can use combinations of these functions to see how conditioning can restrict pre-image sets to create statistical independence.

Let $X := r_1 + r_2$ and $Y := r_1 r_3$. These are coarse-graining functions on $[0, 1)$ with values in $\{0, 1, 2\}$ and $\{0, 1\}$, respectively. It is straightforward to check that they are not statistically independent (preferably by sketching their graphs and the relevant pre-images). For instance, we have

$$\lambda(\{x : X(x) = 0, Y(x) = 0\}) = \frac{1}{4} \neq \lambda(\{x : X(x) = 0\})\lambda(\{x : Y(x) = 0\}) = \frac{1}{4} \cdot \frac{6}{8}.$$

Here, λ is used as in Chap. 3 to denote the Lebesgue measure on $[0, 1]$. It is easy to identify r_1 as the "common cause" or "local explanation" for the correlation between X and Y. We can conditionalize on a value of r_1, let's say $r_1 = 0$. This restricts the pre-image set $\{x : X(x) = 0, Y(x) = 0\}$ to $[0, 1/2)$. And indeed, on this set, we now obtain statistical independence. More precisely, we have

$$\lambda_c(X = 0) := \lambda(X = 0 | r_1 = 0) = \frac{\lambda(\{x : X(x) = 0\} \cap \{x : r_1(x) = 0\})}{\lambda(\{x : r_1(x) = 0\})}$$

$$= \frac{\lambda([0, 1/4))}{\lambda([0, 1/2))} = \frac{1}{4} \cdot \frac{2}{1} = \frac{1}{2},$$

$$\lambda_c(Y = 0) := \lambda(Y = 0 | r_1 = 0) = \frac{\lambda(\{x : Y(x) = 0\} \cap \{x : r_1(x) = 0\})}{\lambda(\{x : r_1(x) = 0\})}$$

$$= \frac{\lambda([0, 1/2))}{\lambda([0, 1/2))} = 1,$$

and for the joint conditional distribution

$$\lambda_c(X = 0, Y = 0) = \frac{\lambda(\{x : X(x) = 0, Y(x) = 0\} \cap \{x : r_1(x) = 0\})}{\lambda(\{x : r_1(x) = 0\})}$$

$$= \frac{\lambda([0, 1/4))}{\lambda([0, 1/2))} = \frac{1}{2}$$

$$= \frac{1}{2} \cdot 1 = \lambda_c(X = 0)\lambda_c(Y = 0).$$

To prove the statistical independence of X and Y under the conditional measure λ_c, we must, of course, check that the measure factorizes for all possible combinations of values, but this is done analogously and can be carried out as an exercise.

With these examples in mind, we return to the actual issue at hand, namely the correlations observed in the EPRB experiment. Here, the physical theory has to tell us what the relevant variables are that could feature in a local explanation of the correlations (10.11). The correlations will be called *locally explainable* (by the proposed candidate theory) if

$$\mathbb{P}(A, B \mid a, b, \lambda) = \mathbb{P}(A \mid a, \lambda) \cdot \mathbb{P}(B \mid b, \lambda), \tag{10.12}$$

where λ now encodes all physical quantities and events which, according to our theory, could be relevant to the measurement statistics. These variables could describe particles, or fields, or strings, or wave functions; they could include conserved quantities or random variables—whatever the theory has to offer.[8]

Naturally, we have to recover the original probabilities—matching the observed frequencies—when averaging over λ, i.e.,

$$\int_\Lambda \mathbb{P}(A, B \mid a, b, \lambda) \, d\mathbb{P}(\lambda) = \mathbb{P}(A, B \mid a, b), \tag{10.13}$$

where the integral goes over the entire range Λ of the variables λ. In other words, the values of λ can vary in each run of the experiment, but the distribution $\mathbb{P}(\lambda)$ must be such as to produce the correct outcome statistics over multiple runs.

We thus arrive at the following precise definition of locality/nonlocality:

Definition 10.1 A theory is said to be *nonlocal* (in the sense of Bell and EPR), if it predicts correlations between spacelike separated events which are *not* locally explainable within the theory, in the sense of Eq. (10.12).

We can see that (10.12) captures the physical concept of locality as a necessary criterion as follows (consider also Fig. 10.2). From the definition of conditional probabilities, we get

$$\mathbb{P}(A, B \mid a, b, \lambda) = \mathbb{P}(A \mid B, a, b, \lambda) \, \mathbb{P}(B \mid a, b, \lambda). \tag{10.14}$$

Recall that λ is supposed to encode a complete description of the physical state in the past, including in particular all possible "common causes" for the outcomes A and B. In a local theory, therefore, the additional specification of the outcome B and the control parameter b must be redundant for the predictions A. This is expressed by $\mathbb{P}(A \mid B, a, b, \lambda) = \mathbb{P}(A \mid a, \lambda)$. Analogously, the probability of B, conditioned on λ, must no longer depend on the choice of a. Thus $\mathbb{P}(B \mid a, b, \lambda) = \mathbb{P}(B \mid b, \lambda)$. In a local description, the mathematical identity (10.14) must therefore reduce to (10.12).

Remark 10.2 (On the Notion of "Action at a Distance") Previously, we described nonlocality in terms of "action at a distance". The term "action at a distance" is in the right spirit, but somewhat problematic. On the one hand, because it is tainted by

---

[8]Note that the probabilities in (10.12) could all be 1 or 0, which would be the case for deterministic theories in which a complete state description λ uniquely determines the measurement outcomes.

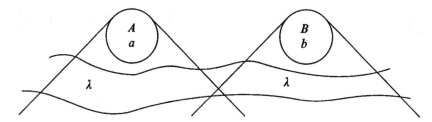

Fig. 10.2 Spacetime diagram of the EPRB experiment. *Diagonal lines* indicate the past light cones and the measurement events occur in spacelike separated regions. λ encodes all relevant factors in the past of the measurement events that could serve to "screen off" the correlations. Source: J.S. Bell, *Speakable and Unspeakable in Quantum Mechanics*, 2nd edn, Cambridge University Press, 2004

Einstein's polemics ("ghost fields", "spooky action at a distance"), and on the other because it may invoke causal intuitions that could turn out to be inappropriate. It is best to leave all further intuitions and questions about "cause" and "effect" aside, and understand nonlocality first and foremost in the sense of the above definition (given by John Bell), that is, in terms of locally inexplicable correlations between distant events.

We now obtain the first important conclusion simply by applying the above definition to standard quantum mechanics. In this case, λ will include the wave function ψ—which is supposed to provide the complete microscopic description of the EPR pair—and possibly also "classical" macroscopic variables $X_1, \ldots, X_n$ describing the particle source, the null state of the detectors, etc. However, $\lambda_{QM} = (\psi, X_1, \ldots, X_n)$ is not sufficient to "screen off" the correlations between the measured particle spins of the EPR pair. The predictions of the theory are still (10.10), so

$$\mathbb{P}(A, B \mid a, b, \lambda_{QM}) \neq \mathbb{P}(A \mid a, \lambda_{QM}) \cdot \mathbb{P}(B \mid b, \lambda_{QM}).$$

In other words, standard quantum mechanics predicts the EPR correlations between distant measurement events without providing sufficient resources for a local explanation. The theory is therefore nonlocal. There is nothing to debate here. We merely have to check the definition.

We thus arrive once again at the EPR dilemma. Standard quantum mechanics is either incomplete or nonlocal. And the question is once again whether it is possible to complete quantum mechanics—or even replace it with an entirely new theory—in order to provide a local explanation of the EPR correlations.

The only additional requirement we shall impose on such a candidate local theory is that the relevant physical variables encoded in λ are independent of the control parameters a and b. Put simply, this is the assumption that the orientations of the Stern–Gerlach magnets can be freely (or randomly) chosen in the experiment. Formally,

$$\mathbb{P}(\lambda \mid a, b) = \mathbb{P}(\lambda) . \tag{10.15}$$

A local explanation violating this assumption is said to be *conspiratorial*. Why conspiratorial? Well, there are basically two reasons why (10.15) could be violated. We either have a form of *retro-causation* in the sense that the parameter choices a and b in the future influence the physical state λ in the past, or we have a form of *superdeterminism* in which, whatever determines the outcomes of the measurements also determines what measurements will be performed in the first place. Such a theory would threaten to render absurd the entire scientific enterprise, which is based on the belief—be it only a stubborn illusion—that we can freely decide what to probe in experiments, i.e., that Nature is not like a dictatorial regime guiding investigators through controlled tours that present a distorted view of reality.

That said, we don't have to go into the difficult subject of "free will" to see why superdeterminism is absurd. In practice, the parameter choices a and b are usually made by some sort of (quantum) random number generator. In one experiment by Shalm et al., the choices were made to depend, in addition, on bitmaps generated from various sources including the 1985 movie *Back to the Future*. In principle, they could also be made by fluctuations in the stock market or radio signals from distant galaxies. The idea that none of these physical processes could be treated as independent from the initial state of the EPR particles is much more mind-boggling and spooky than any action at a distance. Such conspiracies are therefore excluded in the following.

From these two assumptions, the *locality assumption* (10.12) and the *no conspiracy assumption* (10.15), we can derive the so-called CHSH inequality:

Theorem 10.1 (CHSH Inequality) *For fixed orientations a, b, we consider the expected value of the product $A \cdot B$, that is*

$$E(a, b) := \mathbb{E}(A \cdot B \mid a, b) , \tag{10.16}$$

where

$$\mathbb{E}(A \cdot B \mid a, b) = \mathbb{P}(A = +1, B = +1 \mid a, b) + \mathbb{P}(A = -1, B = -1 \mid a, b) \tag{10.17}$$
$$-\mathbb{P}(A = +1, B = -1 \mid a, b) - \mathbb{P}(A = -1, B = +1 \mid a, b) .$$

Under the assumptions (10.12) (locality) and (10.15) (no conspiracy) the CHSH *inequality*

$$S := |E(a, b) - E(a, b')| + |E(a', b) + E(a', b')| \leq 2 \tag{10.18}$$

then holds for any four parameter values a, a', b, b' and any distribution of λ.

The proof is quite simple and will be given in Sect. 10.4.1. Using (10.6) or (10.10) in (10.17), the quantum mechanical predictions then yield $E(a, b) = -\cos(\sphericalangle(a, b))$. The maximal violation of the CSHS inequality occurs for $a = 90°$, $a' = 0°$, $b = 45°$, $b' = -45°$, when we obtain

$$S = 2\sqrt{2}, \tag{10.19}$$

which is clearly greater than 2.

Various experiments confirm the violation of the CHSH inequality (10.18) and thus the impossibility of a local (non-conspiratorial) account of the EPR correlations. Recent measurements of electron spins on EPR pairs over a distance of 1.3 km obtained a value of $S = 2.42 \pm 0.20$ and thus a significant violation of the CHSH inequality, consistent with the predictions of quantum mechanics.[9] This experiment by Hensen et al. is considered to be the first "loophole free" test of the CHSH inequality (others have already followed since). This means, in particular, that all non-superluminal influences between the measurements events (including the random choices of the control parameters) have been excluded, and that the detector efficiency was high enough to rule out the possibility that the statistics could be skewed by those particles that have never been registered. All in all, the experimental evidence is considered to be conclusive, and all the evidence comes down on the side of nonlocality.[10]

Let us sum up one last time. The *only* assumptions underlying the derivation of (10.18) are the locality assumption (10.12) and the no-conspiracy assumption (10.15). The relevant experiments observe a significant violation of this CHSH inequality, thus excluding all local, non-conspiratorial explanations of the EPR correlations. Hence, Nature is indeed "conspiring" against us, or—and this is really the only serious option—Nature is nonlocal.

10.4.1 Derivation of the CHSH Inequality

We consider the expected value (10.17):

$$E(a, b) = \mathbb{P}(A = +1, B = +1 \mid a, b) + \mathbb{P}(A = -1, B = -1 \mid a, b) \tag{10.20}$$

$$-\mathbb{P}(A = +1, B = -1 \mid a, b) - \mathbb{P}(A = -1, B = +1 \mid a, b).$$

---

[9]B. Hensen et al., Loophole-free Bell inequality violation using electron spins separated by 1.3 kilometres. Nature **526**, 682–686 (2015).

[10]Since the relevant experiments exclude only influences that propagate at most with the speed of light, we could ask, at least as long as we think non-relativistically, whether nonlocal influences have to be truly *instantaneous* or whether they could act with a finite (though superluminal) velocity. The answer is that they could not, at least not if superluminal communication is excluded; see N. Gisin, Quantum correlations in Newtonian space and time. In: D. Struppa and J. Tollaksen (Eds.), *Quantum Theory: A Two-Time Success Story*, Springer, 2014.

With assumptions (10.12) and (10.15), we can write this as

$$E(a, b) = \int_\Lambda \Big[\mathbb{P}(A = +1 \mid a, \lambda) - \mathbb{P}(A = -1 \mid a, \lambda) \Big]$$
$$\times \Big[\mathbb{P}(B = +1 \mid b, \lambda) - \mathbb{P}(B = -1 \mid b, \lambda) \Big] d\mathbb{P}(\lambda) \,.$$

Here, we have used the fact that, according to (10.12), the joint probabilities factorize after conditionalizing on λ. Expanding the brackets and averaging over λ, we thus get back (10.17). The expressions inside the brackets are nothing else than the expectations of A and B conditionalized on λ. With the abbreviations

$$\overline{A}(a, \lambda) := \mathbb{P}(A = +1 \mid a, \lambda) - \mathbb{P}(A = -1 \mid a, \lambda) \,,$$
$$\overline{B}(a, \lambda) := \mathbb{P}(B = +1 \mid b, \lambda) - \mathbb{P}(B = -1 \mid b, \lambda) \,,$$

we thus get

$$E(a, b) = \int_\Lambda \overline{A}(a, \lambda) \overline{B}(b, \lambda) \, d\mathbb{P}(\lambda) \,,$$

where

$$|\overline{A}(a, \lambda)| \leq 1 \,, \quad |\overline{B}(b, \lambda)| \leq 1 \,. \tag{10.21}$$

Now we add/subtract the correlations for the different orientations a, a', b, b'. To begin with, we have

$$E(a, b) - E(a, b') = \int_\Lambda \overline{A}(a, \lambda) \Big[\overline{B}(b, \lambda) - \overline{B}(b', \lambda) \Big] d\mathbb{P}(\lambda) \,, \tag{10.22}$$

whence (10.21) implies

$$|E(a, b) - E(a, b')| \leq \int_\Lambda \Big| \overline{B}(b, \lambda) - \overline{B}(b', \lambda) \Big| d\mathbb{P}(\lambda) \,. \tag{10.23}$$

Then, analogously,

$$E(a', b) + E(a', b') = \int_\Lambda \overline{A}(a', \lambda) \Big[\overline{B}(b, \lambda) + \overline{B}(b', \lambda) \Big] d\mathbb{P}(\lambda) \,, \tag{10.24}$$

and (10.21) implies

$$|E(a', b) + E(a', b')| \leq \int_{\Lambda} \left|\overline{B}(b, \lambda) + \overline{B}(b', \lambda)\right| d\mathbb{P}(\lambda). \qquad (10.25)$$

Adding (10.23) and (10.25) thus yields

$$|E(a, b) - E(a, b')| + |E(a', b) + E(a', b')| \qquad (10.26)$$
$$\leq \int_{\Lambda} \left|\overline{B}(b, \lambda) - \overline{B}(b', \lambda)\right| + \left|\overline{B}(b, \lambda) + \overline{B}(b', \lambda)\right| d\mathbb{P}(\lambda).$$

Now we use the following simple inequality: if $|x|, |y| \leq 1$, then $|x-y|+|x+y| \leq 2$. This is easy to see by squaring the last expression. We get

$$(|x - y| + |x + y|)^2 = 2x^2 + 2y^2 + 2|x^2 - y^2|,$$

which is equal to $4x^2$ (if $x^2 \geq y^2$) or $4y^2$ (if $x^2 < y^2$), so it is in every case less than or equal to 4. Hence, we have

$$|\overline{B}(b, \lambda) - \overline{B}(b', \lambda)| + |\overline{B}(b, \lambda) + \overline{B}(b', \lambda)| \leq 2,$$

and together with (10.26), we obtain the CHSH inequality

$$|E(a, b) - E(a, b')| + |E(a', b) + E(a', b')| \leq 2.$$

10.5 Nonlocality and Faster-than-Light Signaling

A plausible concern is that the nonlocality of quantum mechanics might be incompatible with Einsteinian relativity. So far, we have only considered non-relativistic quantum theories, but the existence of nonlocal correlations is an empirical fact, and if that fact made any relativistic account impossible, one of the pillars of modern physics would be shattered.

There is indeed a certain tension between nonlocality and relativity, but it is not a straight-up contradiction. This tension and possible ways to resolve it will be discussed in more detail in Chap. 11. There is one major concern, though, that we shall address right away: it would be natural to think that the nonlocal correlations of quantum mechanics could be exploited for superluminal signaling, i.e., faster-than-light communication. Fortunately, this turns out to be impossible.

Why are superluminal signals so problematic? Common wisdom is that they are explicitly excluded by special relativity, though this statement is oversimplified and requires further elaboration, if only because the notion of "signal" is not so very precise. The fundamental observation is that, in relativistic spacetime, there is no absolute temporal order between spacelike separated events. If A and B are two events such that B lies outside the light cone of A and vice versa, then there are some coordinate frames in which A occurs before B and some in which B occurs before A (and indeed some in which the two events occur at the same time). The worry is now that superluminal signaling could lead to causal paradoxes.

Consider the following hypothetical scenario. Candidate A is competing in a game show and must chose between three doors, one of which contains the jackpot. She chooses door number 1 which turns out to be a loser. However, sneaky as she is, she has devised a plan. She has prepared an entangled quantum system to send a superluminal signal to her accomplice B with the information that the prize is not behind door number 1. In the reference frame of A, the signal arrives more or less instantaneously, that is, the delay can be made arbitrarily small.

Meanwhile, B is in a spaceship moving at a constant speed close to the speed of light. Relative to her rest frame, the message from A comes from the future! Now, by sending a superluminal signal back to A, she can warn her not to pick door number 1 before A makes her choice in the game show. So this time, A, having received the tip "from the future", chooses door number 3 and takes home the jackpot.

But wait a minute. This is not just cheating. The described series of events is logically inconsistent. If A picks door number 3 because B told her to, she never opened door number 1, never had to signal to her accomplice, and never received the tip in the first place. So we had better make sure that our physical theory does not allow such signaling schemes.

Consider the EPRB experiment from the point of view of Bohmian mechanics. Let us assume that the first experimenter A could know the exact positions of the entangled particles at time $t = 0$, when the system is prepared, and thus predict the exact measurement outcomes from the deterministic Bohmian laws. She could thus agree with her colleague B on the following communication protocol. The Stern–Gerlach apparatus of B always remains oriented in the z-direction. Thus, if A measures the spin of her particle in the x-direction, B will register "spin UP" or "spin DOWN" with equal probability $1/2$. To send a signal, however, A can decide to measure the z-spin of her particle if and only if she knows that the outcome will be "spin UP" (otherwise, she keeps measuring the x-spin). This increases the probability of B measuring "spin DOWN" from $1/2$ to $3/4$. Hence, with a sufficient number of measurements, B can determine with great confidence whether A is signaling or not. And with a binary code of "signal" or "no signal" (1 or 0), the two could exchange arbitrarily complex messages faster than the speed of light.

The impossibility if this superluminal communication follows from the theorem of *absolute uncertainty*, proved in Sect. 4.2 as a consequence of quantum equilibrium. Experimenter A cannot know the particle positions and thus the outcome of the spin measurements with greater accuracy than corresponding to the Born rule. In the present case, with an EPR pair prepared in the spin singlet state, this means that A cannot predict more about the spin measurements than that they will yield "spin UP" and "spin DOWN" with probability $1/2$.

In general, absolute uncertainty implies that the experimenter cannot reliably choose her parameter settings in any way that leads to different outcome statistics than those predicted by quantum mechanics. In this sense, i.e., in the sense of Born's rule, the outcomes of the spin measurements are random. And by averaging over the possible outcomes of A's measurement (here, $A = +1$ and $A = -1$), we obtain for the marginal distribution of the distant measurement event B

$$\sum_{A=\pm 1} \mathbb{P}(B \mid A, a, b) \mathbb{P}(A \mid a) = \mathbb{P}(B \mid b), \qquad (10.27)$$

which is independent of the parameter a. This means that A cannot influence the measurement statistics of B, and, of course, vice versa. Equation (10.27) is therefore also known as the *no signaling condition*. For quantum mechanical correlations, this condition is always satisfied, even for entangled systems that violate the locality assumption (10.12), like those used to derive the Bell or the CHSH inequality (10.18).

In the context of the GRW theory the status of (10.27) is even more evident. Here, the measurement outcomes A and B are intrinsically random (and only determined through the random collapse process), so there's nothing the experimenters could know even in principle that would allow them to predict these outcomes with higher accuracy. In the Many Worlds theory, (10.27) holds within individual branches to the extent that Born's rule does. From the more holistic point of view, it is impossible to influence measurement outcomes anyway. Whatever the experimenter may do, the outcome of the spin measurements will always be "spin UP" *and* "spin DOWN".

In conclusion, quantum nonlocality should be first and foremost understood according to Definition 10.1, that is, in terms of locally inexplicable correlations (see also the subsequent Remark 10.2). "Signaling", however, requires a sufficient degree of (agential) control over such correlations—and all quantum theories agree that this is impossible.

Superluminal communication is often made out as the bogeyman when discussing alleged unphysical consequences of nonlocal interactions. These concerns, however, are based on certain intuitions about free will and human intervention that would result in causal paradoxes in hypothetical situations. It is also conceivable—though maybe not very plausible—that nature allows for violations of relativistic causality while some cosmic principle (maybe by enforcing particular boundary conditions) excludes paradoxical solutions. In terms of our previous example, this could mean that the laws of nature only allow for solutions in which candidate A picks the right door from the very beginning, e.g., because her accomplice B told her in advance that the prize was behind door number 3, which B knew because A signalled it after opening door number 3. This sequence of events also describes a *causal loop*, but in this case it is logically consistent.

In the relativistic context, there is, however, another (better) reason to insist on the no signaling condition (10.27). We recall that, in relativistic spacetime, there is no absolute temporal order between two spacelike separated events A and B. In some reference frames, A occurs before B, and in other reference frames B occurs before A (and in some particular frames, the two events are simultaneous). In order for the measurement statistics to be consistent with a relativistic (Lorentz covariant) description, they cannot therefore depend on any *order* in which spacelike separated measurements are performed. We already know what this means formally from Chap. 7: the operators associated with such measurements (and hence their spectral decompositions) must commute. Then and only then will we find, for arbitrary wave function ψ and measurement outcomes A, B, that

$$\mathbb{P}(A, B) = \langle \psi \mid P_A P_B \mid \psi \rangle = \langle \psi \mid P_B P_A \mid \psi \rangle, \tag{10.28}$$

independently of the order of measurement. From the commutator relations

$$[P_A, P_B] = 0 \tag{10.29}$$

of the associated projections, the no signaling condition (10.27) can be readily derived. For two commuting spectral decompositions $(P_A^a)_A$ and $(P_B^b)_B$ (here indexed by the control parameters a and b), we have

$$\sum_A \mathbb{P}(B \mid A, a, b)\,\mathbb{P}(A \mid a) = \sum_A \frac{\langle \psi \mid P_B^b P_A^a \mid \psi \rangle}{\langle \psi \mid P_A^a \mid \psi \rangle} \langle \psi \mid P_A^a \mid \psi \rangle = \sum_A \langle \psi \mid P_B^b P_A^a \mid \psi \rangle$$

$$= \sum_A \langle \psi \mid P_A^a P_B^b \mid \psi \rangle = \langle \psi \mid \Big(\sum_A P_A^a \Big) P_B^b P_B^b \mid \psi \rangle$$

$$= \langle \psi \mid P_B^b \mid \psi \rangle = \mathbb{P}(B \mid b),$$

where we have used the fact that $\left(\sum_A P_A^a \right)$ is the identity operator.

In the context of the EPRB experiment, it is easy to check that the spin observables commute since the operators act on only one tensor component of the entangled wave function. That is,

$$\left(\sigma_a^{(1)} \otimes 1 \right)\left(1 \otimes \sigma_b^{(2)} \right) = \sigma_a^{(1)} \otimes \sigma_b^{(2)} = \left(1 \otimes \sigma_b^{(2)} \right)\left(\sigma_a^{(1)} \otimes 1 \right).$$

A final warning, however. For the reasons just explained, relativistic quantum (field) theories generally postulate the commutator relation (10.29) for observables or field operators associated with spacelike separated regions of spacetime. In the literature, this often goes by the misleading name "locality condition", since it is a condition on local operators. However, as we just saw, (10.29) has little to do with locality in the sense of Bell—which is also violated in relativistic quantum theories, and if it wasn't, these theories would be wrong—although it does guarantee the relativistic consistency of measurement statistics and, as proven above, the impossibility of faster-than-light signaling.

Relativistic Quantum Theory

11

I don't think we have a completely satisfactory relativistic quantum-mechanical model, even one that doesn't agree with nature, but, at least, agrees with the logic that the sum of probability of all alternatives has to be 100%. Therefore, I think that the renormalization theory is simply a way to sweep the difficulties of the divergences of electrodynamics under the rug. I am, of course, not sure of that.

Richard P. Feynman, Nobel Lecture 1965[1]

This will be a very hard chapter. Not because of abstract and technically advanced mathematics, which can easily be learned as soon as the underlying physics is clear, that is, as soon as the need for abstraction is evident. It will be hard for two reasons. First, there does not exist a fundamental, mathematically coherent and consistent formulation of a relativistic quantum theory with interaction that could extend the analysis of the foregoing chapters to relativistic physics. We shall say more about that in a moment.

The second reason is the tension between the nonlocality of Nature, which we have talked about, and relativity. It is general folklore that, according to relativistic physics, signals cannot be sent faster than the speed of light. This is too easily misread as: interactions can at best happen with the speed of light so we're in for trouble, given the nonlocality of Nature. But how, then, can these two features be rendered compatible?

As already discussed in Sect. 10.5, quantum mechanical nonlocality cannot be used for superluminal communication. In fact, interactions are fundamental, while transmissions of signals in the sense of relativity theory belong to another category, viz., a thermodynamic one. Nevertheless it seems that the nonlocal correlations in the sense of Bell need some kind of synchronisation or absolute simultaneity which would go against the fundamental principles of relativity. Hence there does exist a

---

[1] Online version: http://www.nobelprize.org/nobel_prizes/physics/laureates/1965/feynman-lecture.html

© Springer Nature Switzerland AG 2020

D. Dürr, D. Lazarovici, *Understanding Quantum Mechanics*,

https://doi.org/10.1007/978-3-030-40068-2_11

tension between relativity and nonlocality.[2] The aim of this and the next chapter will be to understand this better.

But first we wish to explain our claim that a good fundamental relativistic quantum theory is still lacking, where "good" means a formulation which is as mathematically clean and consistent as Schrödinger's equation is for non-relativistic physics. That claim may look at first sight as though it goes too far, since courses on quantum field theory are standard courses in the physics curriculum and usually taken as synonymous with relativistic quantum physics. Moreover and in particular, the so-called Standard Model of particle physics is a quantum field theory which is empirically enormously successful. So what exactly do we have in mind?

11.1   Difficulties of "First" and "Second" Class

In 1963, the journal *Scientific American* published a famous and since then often cited essay by Paul A.M. Dirac with the title *The evolution of the physicist's picture of nature*. In this, he described the actual state of quantum theory and in particular of relativistic quantum theory, pointing out the most important open problems. These he separated into "class one difficulties" and "class two difficulties", later referred to by Bell as "first class" and "second class". The first class difficulties are largely those we have discussed so far in this book: the measurement problem or the missing ontology in quantum mechanics. About those, Dirac says that their solutions will have to wait until the second class difficulties have been solved.

11.1.1 Infinite Mass

The second class difficulties are first of all technical and refer to the formulations of relativistic field theories, which contain in general infinite terms, something like a division by zero which we know is mathematically problematic. The first such difficulty is a well known nuisance which already appears in the first relativistic theory, namely, Maxwell–Lorentz electromagnetism. It is called *ultraviolet divergence*, because it has to do with ultra high energies (and hence short length scales) and because high energy radiation begins in the ultraviolet spectrum. In order to comfort us, it is often said that this divergence shouldn't come as a surprise, because the Maxwell–Lorentz theory is only a low energy approximation to a fundamental theory (still to be found), in the sense that situations corresponding to low energies can be dealt with using Maxwell–Lorentz theory. These can be situations in which the fundamental nature of the elementary charges plays no role, as when electromagnetism is used in engineering practice.

---

[2]See T. Maudlin, *Quantum Non-Locality and Relativity*, 3rd edn, Wiley-Blackwell, 2011.

Be that as it may, Dirac's hope for a mathematically consistent theory has still not been fulfilled and we still have to deal with ultraviolet divergencies—also in quantum field theory. We wish to discuss briefly the simplest example. It has to do with the Coulomb field of an electrical point charge, for example, an electron.[3] The Coulomb field energy, which runs along with the electron with charge e, diverges at the position of the electron, let's say at $r = 0$, where it goes as $F \sim e^2/r$. According to the relativistic mass–energy equivalence $E = m_F c^2$, the field itself adds the divergent contribution

$$m_F \sim \lim_{r \to 0} \frac{e^2}{rc^2} = \infty$$

to the mass of the electron. This is also referred to as the electromagnetic mass of the electron arising from the "self-energy". Hence the electron, which is inseparably connected to its own Coulomb field, would have an infinite mass, which contradicts our experience. One idea here is to say that the mass m of the electron has two parts, the "bare" mass m_0 and added to that the field mass m_F, whence $m = m_0 + m_F$. Since the measured mass m of the electron is finite and $m_F = +\infty$, we must have $m_0 = -\infty + m$, i.e., the bare mass is negatively infinite in exactly such a way that the sum yields the measured mass.

But we know that energy can be gauged so that we can simply subtract a (possibly infinite) constant and consider the difference as the physical energy. In this sense we can arrive at the idea of just gauging the self-energy of the electron away, which is more or less the same as saying that the electron does not interact with its own field. Unfortunately, that does not work very well, because the field energy shows up anyway as inertial mass in the dynamics of the electron: when an electron accelerates, it radiates away, i.e., loses, electromagnetic energy. That loss appears as a drag on the motion of the charge. If we compute this drag, which is due to the action of the radiation on the electron itself (referred to as a self-interaction), we find the self-energy reappearing in a contribution to the inertial mass.

In quantum theory, this friction effect is still present. It is manifested as a broadening of the spectral lines in the so-called Lamb shift, whose computation uses the trick of writing $m_0 = -\infty + m$. This way of handling the divergence is an example of what is known as *renormalization*, a way of getting from infinite terms to finite ones which can then be gauged by the measured values (like the measured electron mass). Renormalization allows very precise predictions and has made quantum physics an empirically well-confirmed theory in recent decades. Indeed, when we talk about quantum field theories these days, we really mean the theory *including* these adjustments which allow us to get rid of the infinities in a very clever way and which thus lead to predictions for experiments.

---

[3] We do not wish to further scrutinize the assumption of a point particle. It is sufficient here to understand that a point has a relativistically invariant form. There are other relativistically invariant forms, but so far considerations along such lines have not led to empirically adequate formulations.

11.1.2 Infinite Pair Creation

There is another infinity which may occur, namely an infinity of particles. This is also associated with Dirac, who had been looking for a relativistic Schrödinger equation that would induce a timelike four-current $j^\mu, \mu = 0, 1, 2, 3$, i.e., the generalisation of the quantum flux (1.5), which when viewed as a four-current $j = (j^0, \mathbf{j})$ has the zero-component $j^0 = \rho = |\psi|^2$. Before formulating his famous Schrödinger equation, which lives in a Galilean spacetime, Schrödinger had written down a relativistic equation which later became known as the Klein–Gordon equation. This is a second order wave equation in the space and time variables and it has the disadvantage that it does not in general allow for a current which is future-oriented, i.e., it sometimes points towards the past.

In any case, Dirac managed to write down a relativistic wave equation which allows a timelike current. It is known as the *Dirac equation* for the electron. In a form which is not manifestly relativistic, and with an external electromagnetic field, the Dirac equation reads

$$
\mathrm{i}\frac{\hbar}{mc^2}\frac{\partial\psi(t,\mathbf{x})}{\partial t} = -\sum_{k=1}^{3}\alpha_k\left[\mathrm{i}\frac{\hbar}{mc}\partial_k + \frac{e}{mc}A_k(t,\mathbf{x})\right]\psi(t,\mathbf{x}) + \left(\frac{e}{mc^2}A_0\mathbf{E} + \beta\right)\psi(t,\mathbf{x})
$$

$$
\equiv \left[D^0 + \tilde{A}(t,\mathbf{x})\right]\psi(t,\mathbf{x}). \tag{11.1}
$$

Let us just explain the notation. To begin with, t and $\mathbf{x} = (x_1, x_2, x_3)$ denote the time and space coordinates in a given reference frame, and $\partial_k = \partial/\partial_{x_k}$ is shorthand for the partial derivatives. m and e are the (measured) mass and charge of the electron and c is the speed of light. α_k are 4×4-matrices built using the Pauli matrices, $\mathbf{E}$ is the 4×4 unit matrix, and

$$
\beta = \begin{pmatrix} 1 & 0 & 0 & 0 \\ 0 & 1 & 0 & 0 \\ 0 & 0 & -1 & 0 \\ 0 & 0 & 0 & -1 \end{pmatrix}.
$$

Finally, $A = (A_0, A_1, A_2, A_3) = (A_0, \mathbf{A})$ is the four-potential, known from electrodynamics, whose derivatives yield the electric and magnetic fields. For example, the magnetic field is given by $\mathbf{B} = \nabla \times \mathbf{A}$. In the second line of (11.1), D^0 stands for the free Dirac operator without external field. ψ is no longer a complex-valued function but a spinor, although it differs from the spinors appearing in the Pauli equation (1.28), because it now has four components. Roughly speaking, we may think of the Dirac spinor as built from two Pauli spinors.

The advantage of the above (non-manifestly relativistic) way of writing the Dirac equation is that, by analogy with the Schrödinger case, we can now read the operator on the right-hand side of (11.1) as the energy operator. The exact values of the matrix components are not important for our purpose. All we shall need for later is that they satisfy the commutation relations

$$\alpha_j \beta = -\beta \alpha_j . \tag{11.2}$$

The conserved four-current has the form

$$j^\mu = \left((\psi, \psi), (\psi, \alpha_1 \psi), (\psi, \alpha_2 \psi), (\psi, \alpha_3 \psi) \right),$$

where $(\cdot, \cdot)$ denotes the spinor scalar product, as in (1.29), and the positivity of the zero component (ψ, ψ) shows that the current is timelike.

Much more could be said about the Dirac equation, for example, that it can be viewed as a kind of "square root" of the Klein–Gordon equation, or that we can understand the four-potential **A** geometrically as arising from a "covariant derivative", or again that the Dirac spinors can be viewed as carrying a representation of the Lorentz group. However, we won't need any of that for our present purposes, and moreover these things can be found in many good textbooks.

Remark 11.1 (Dirac Equation in Relativistic Notation) To prepare for later discussions, we write the Dirac equation in manifest Lorentz invariant notation. To do so, we first define the Dirac matrices $\gamma^0 := \beta$, $\gamma^k := \beta \alpha_k$, $k = 1, 2, 3$, which are built from the matrices α and β mentioned above. By virtue of (11.2), we have the following commutation relations:

$$\{\gamma^\mu, \gamma^\nu\} = \gamma^\mu \gamma^\nu + \gamma^\nu \gamma^\mu = 2\eta^{\mu\nu} , \tag{11.3}$$

where $\eta^{\mu\nu} = \mathrm{diag}(1, -1, -1, -1)$ denotes the Minkowski metric. With these new matrices we can bring the Dirac equation into the simple form

$$\left[\gamma^\mu (i\partial_\mu - eA_\mu) - m \right] \psi = 0 , \tag{11.4}$$

where we sum over repeated indices (here $\mu = 0, 1, 2, 3$) and where the units are now chosen such that the speed of light and Planck's constant are equal to 1. The four-current can be written compactly as

$$j^\mu = \overline{\psi} \gamma^\mu \psi , \tag{11.5}$$

with $\overline{\psi} := \psi^+ \gamma^0$, and where ψ^+ is the transposed complex conjugate of ψ. The Dirac equation implies that the current is conserved, i.e., its four-divergence is zero, viz.,

$$\partial_\mu j^\mu = 0 . \tag{11.6}$$

If we write the four current with respect to a special frame of reference as $j^{\mu} = (\rho, \mathbf{j})$, we recognize the generalization of the quantum flux equation $\partial_t \rho + \text{div}\,\mathbf{j} = 0$.

Let us now turn to the energy spectrum and note that, in the rest frame of the electron, i.e., when its momentum is zero, and if we assume that the field A is also zero, we can drop the terms in α in the Dirac equation. The eigenvalue equation for the rest energy is then

$$E\psi = mc^2 \beta \psi ,$$

where β has eigenvalues $1, 1, -1, -1$ with canonical eigenvectors $\mathbf{e}_n, n = 1, 2, 3, 4$. If we change to a moving frame of reference, all values in $(-\infty, -mc^2) \cup (mc^2, +\infty)$ can become energy values.[4]

This is good and at the same time bad. It is bad, because an energy spectrum which is unbounded from below will allow the electron to radiate an unlimited amount of energy. That would clearly solve all energy problems in the world, but it is of course unphysical. The above result is good, because reflection on this problem led Paul Dirac to describe a phenomenon which would forever change our view of the world. That phenomenon was *pair creation and pair annihilation*. Dirac's idea was that, by Pauli's exclusion principle, an electron would not be able to fall to arbitrarily low energy values if the negative energy states were already occupied, i.e., an electron can only sink into the negative energy spectrum if one of the states is empty.[5] The electrons occupying the negative energy spectrum are "invisible" to us, but they can be lifted to positive energies and thus become detectable.

The way the negative energy states of the Dirac equation are filled up is somewhat analogous to the way in which electrons occupy the possible stationary states around an atomic nucleus, except that the charge of the nucleus determines the number of electrons, which then organize themselves according to the Pauli principle, from lower to subsequent higher energy levels. This structure underlies, in particular, the periodic table of elements. Here, we can take an infinite basis, e.g., $\varphi_1, \varphi_2, \ldots$, of the negative energy subspace of the Dirac Hamiltonian and say, in analogy with the atom, that each negative energy state is occupied by one electron (see Sect. 11.3.1 for a more precise explanation). This construction is called the *Dirac sea*. Of course, in contrast to the atom, the Dirac sea contains infinitely many particles and no positively charged nucleus around which the electrons assemble. The negative energy states come from the dynamics of the Dirac equation alone.

---

[4]If the field A is not zero, then there is in general no longer a spectral gap in $(-mc^2, mc^2)$.

[5]One may turn the argument around and take the avoidance of the radiation catastrophe as justification for the Pauli exclusion principle. In fact, along these lines, a generalization of the Pauli exclusion principle was later established, known as the spin–statistics theorem, which states that many-particle half integer spin wave functions must be antisymmetric, i.e., fermionic wave functions (see Remark 4.9).

The Dirac sea is mathematically equivalent to the so-called vacuum of the quantum field theory built on the basis of the Dirac equation, i.e., quantum electrodynamics (QED). It is important to emphasize this mathematical equivalence because it is often claimed that the Dirac sea is an "old and outdated" picture which in the modern literature has been made precise by the notion of the vacuum, suggesting that the sea picture is kind of problematical while the vacuum is unproblematical because it is precise. So let us repeat that, mathematically, the two approaches are equivalent, so if a mathematical problem occurs in one, then it also occurs in the other.[6]

Let us now consider this sea with infinitely many electrons. By the antisymmetry of the wave function, or equivalently, by the Pauli principle, electrons cannot sink into the sea (by analogy with the electrons in an atom), thereby radiating energy. But one big question remains: Why doesn't the sea create problems because of its infinite negative charge and mass? Why don't we feel that? Dirac's opinion, which is the generally accepted one, was that the distribution of the particles in the sea would fill the universe so homogeneously that all possible influences would average out to zero. However, once in a while, in "certain physical situations", we should notice some effects of the Dirac sea, in particular, whenever a particle is lifted from it to a positive energy state ($\geq mc^2$). Then an electron (as we know it) will suddenly appear, along with an unoccupied state or "hole" in the sea. The latter will behave like an electron with positive charge. If we call this hole a *positron*, then we have the phenomenon of *pair creation*. From the hidden sea, there thus appears an electron and a positron. If on the other hand there is a hole in the sea, then an electron can fall into that hole and we have *pair annihilation*—the electron and the hole both disappear. All particles are nevertheless governed by the Dirac equation.

In this construction, however, we have not taken into account the fact that charges interact with each other by sending and receiving electromagnetic radiation. The reason for this omission is that an external electromagnetic field (the field $\mathbf{A}$ in the Dirac equation) already creates a problem of infinity. Indeed, the weakest field $\mathbf{A}$ will suffice to lift infinitely many particles from the sea! How can we see that? By considering the conditions under which an electron can be lifted. We do that by revisiting our simple insight regarding an electron at rest. Here, the negative energy eigenvectors $\mathbf{e}_3$, $\mathbf{e}_4$ will develop positive energy components under the influence of the A-field in the Dirac equation. The latter field appears in combination with the α-matrices according to

$$\frac{e}{mc}\,\boldsymbol{\alpha} \cdot \mathbf{A} := \frac{e}{mc} \sum_{k=1}^{3} \alpha_k A_k(t, \mathbf{x})\,.$$

---

[6]See, e.g., D.-A. Deckert, D. Dürr, F. Merkl, M. Schottenloher, Time-evolution of the external field problem in quantum electrodynamics. J. Math. Phys. **51** (12), 122301 (2010); arXiv:0906.0046, for more details.

Recalling the commutation relation (11.2), we see that $\beta\alpha_j\mathbf{e}_3 = -\alpha_j\beta\mathbf{e}_3 = \alpha_j\mathbf{e}_3$, which means that $\alpha_j\mathbf{e}_3$ becomes an eigenvector of positive energy. In short, $\mathbf{A} \neq 0$ "rotates" negative energy eigenvectors onto positive ones, and the magnitude $|\mathbf{A}|$ of the field doesn't matter. As long as it is not equal to zero, the rotation will occur.[7] And this in turn means that infinitely many pairs will appear. Such a situation can no longer be handled in the formalism of Hilbert spaces (actually Fock spaces, which are Hilbert spaces allowing variable particle numbers) and unitary time evolutions thereon[8] (see Sect. 11.3.2 and the references in footnote 7 for a more detailed discussion of the connection between the Dirac sea and Fock spaces).

However, this infinity of pair creations can be renormalized to get predictions for experiments. For example, measurements are usually made in scattering situations, i.e., situations where the field $\mathbf{A}$ in the Dirac equation is zero at very early and very late times. We can consider a scattering experiment where particles collide and the interaction happens only for a short period of time. This can then be understood as follows: the infinitely many pairs are only present for a short period of time when the interaction is present, and after that everything disappears more or less, back into the Dirac sea, i.e., apart possibly from finitely many created pairs. For this reason, the terminology of *virtual particles* is often used for the infinitely many pairs, because they are only there for a short time to create a bit of mathematical trouble, but they no longer appear in the products of the reaction process.

But mathematics and physics are different facets of the same thing, and if a mathematical description slips out of our hands, we have a problem with our physical picture of the world as well. Moreover, an ontologically complete theory which is supposed to describe the world at every hour of the day and night cannot restrict itself to describing scattering situations alone.

We conclude that the great innovation of relativistic quantum theory is first of all that particles can be created and annihilated—at least, that's the way things appear to us. The Dirac sea picture explains this phenomenon by assuming that all particles have always been and will always be there, but that we observe only those particles which set themselves apart from the homogeneous equilibrium-like distribution— the Dirac sea. The mathematical description, however, is problematic because even weak interactions can lead to an infinite rate of pair creations.

---

[7] For a mathematically precise formulation, see D.-A. Deckert, D. Dürr, F. Merkl, M. Schottenloher, Time-evolution of the external field problem in quantum electrodynamics. J. Math. Phys. **51** (12), 122301 (2010), arXiv:0906.0046; D. Lazarovici, Time evolution in the external field problem of quantum electrodynamics. Thesis, LMU Munich, 2011, online version arXiv:1310.1778; D.-A. Deckert and F. Merkl, External field QED on Cauchy surfaces for varying electromagnetic fields, Commun. Math. Phys. **345** (3), 973–1017 (2016); arXiv:1505.06039.

[8] A general result like this is known in quantum field theory as Haag's theorem. It can be formulated, for example, by stating that free and interacting field operators lead to representations of the canonical commutation relations which are not unitarily equivalent. In the Dirac sea picture, this abstract result can be vividly and physically interpreted in terms of the creation of infinitely many pairs.

Another possibility, suggested by the terminology "quantum field theory", is to replace the particle ontology by a field ontology, where the notion of particle (when we talk, for example, of an electron) merely refers to a particular field configuration.

11.2 Field Ontology: What Exactly Is It?

It is unlikely to be immediately clear what is meant by "field ontology" in the connection with quantum fields. Here we shall allow ourselves to go beyond the scope of this book and talk about something that lies near to hand physically, but which is nevertheless mathematically rather demanding. In this section, the reader will be asked to endure a little mathematical abstraction.

By analogy with quantum mechanics, we can approach the question of the field ontology in the following way. In an N-particle system we have a wave function $\psi(\mathbf{q}_1, \ldots, \mathbf{q}_N, t)$ on the configuration space of the N particles. If we now want to talk instead about field configurations, we can replace the point $(\mathbf{q}_1, \ldots, \mathbf{q}_N)$ in configuration space by a field configuration. What this means is the following. A physical field is first and foremost a function $f : \mathbb{R}^3 \to \mathbb{R}^n$, where $n = 1, 2, \ldots$. For $n = 1$, this is called a scalar field. In contrast, the electric field is a vector-valued field, since it takes values in $\mathbb{R}^3$. In this context, physical fields are thus functions on physical space. They can have other mathematical features as well, such as being differentiable, but for the time being we shall ignore such details. The set of all functions (for a given n) will be denoted by $\mathscr{D}$. This set replaces the N-particle configuration space $\mathbb{R}^{3N}$, i.e., an element $f \in \mathscr{D}$ is a field configuration! The wave function now becomes a "super wave function" $\Psi : \mathscr{D} \to \mathbb{C}^k$, where k captures the spinor nature of the wave function. Instead of $\psi(\mathbf{q}_1, \ldots, \mathbf{q}_N, t)$, with $(\mathbf{q}_1, \ldots, \mathbf{q}_N) \in \mathbb{R}^{3N}$, we now have $\Psi(f, t)$ with $f \in \mathscr{D}$.

Let's pursue this analogy a little further. In the mathematics of quantum mechanics, the generic position of a particle gets associated with a position operator simply by defining

$$\hat{\mathbf{X}}\psi(\mathbf{x}) := \mathbf{x}\psi(\mathbf{x}),$$

which means that $\hat{\mathbf{X}}$ is nothing but a multiplication operator. Accordingly, we now associate with fields f operator-valued fields, i.e., *quantum fields* $\hat{F}(\mathbf{x})$:

$$\hat{F}(\mathbf{x})\Psi(f(\mathbf{x})) := f(\mathbf{x})\Psi(f(\mathbf{x})).$$

We also know [see (1.46)] that the operators satisfy the classical equations of motion. In this sense, the quantum field satisfies the classical field equation—for example the Maxwell equations—after "quantisation". Ignoring the technical details, we can then think about a quantum version of the electromagnetic field.

In Sect. 11.3, we shall talk in more detail about the Fock space associated with the Dirac theory of electrons, which also goes under the name of quantum field theory. However, this immediately raises the question of what field configurations should look like in this case. For example, one might think that a field configuration would represent a particle configuration in some way, e.g., by distinguished "bump configurations" in the field. But that does not work out, because a Dirac super wave function on fields in the above sense does not exist. The description of electrons within Fock space (see Sect. 11.3) is probably the only reasonable one. The reason is that electrons are fermions, i.e., the many-particle wave function must be antisymmetric (see Sect. 4.4), and this antisymmetry conflicts with the representation by super wave functions as described above. Fermions are not conducive to a naive field ontology, while bosons are. One could say that fermions are in fact particles, while "bosons" are merely a manner of speaking.

Nevertheless, bosons are commonly viewed as particles. For example, think of photons, the quanta of electromagnetic fields. A first thought might once again be that we should "see" the bosons in the field configuration as "bumps" in the field. But that is also more or less impossible, as we shall explain later on. For now, let us continue with the field ontology.

A Bohmian field ontology could, for example, be handled in a rather analogous way to a particle ontology. Instead of a vector field on the configuration space, which determines the trajectories of the particles, we now have a vector field on the infinite-dimensional function space $\mathscr{D}$. Instead of the partial derivatives, we now have a functional derivative[9] $\delta/\delta f$, whence the analogy goes as follows. As in (4.2), we write Ψ in polar form

$$\Psi(f, t) = R(f, t)e^{iS(f,t)},$$

then use the phase S to define the evolution of the actual field configuration, i.e., plug the actual field configuration $F(\mathbf{x}, t)$ into the guiding equation

$$\frac{\partial F(\mathbf{x}, t)}{\partial t} = \left.\frac{\delta S(f, t)}{\delta f}\right|_{f=F(\mathbf{x},t)},$$

where the super wave function Ψ satisfies a *functional Schrödinger equation*, which is an infinite-dimensional partial differential equation. The "second class" difficulties would in this case relate to the question of whether expressions like the infinite-dimensional partial differential equation are mathematically well defined in

---

[9]This is a natural generalisation of the directional derivative of functions of many variables, in which the derivative is a linear map of directional vectors to real values (viewed geometrically as the value of a slope on a surface). In short, $\langle \nabla F(\mathbf{x}), h \rangle$ is the derivative of F in the direction h at the point $\mathbf{x}$. In the present case, we have an uncountable set of variables and the directional vector becomes a function, while the scalar product sum is replaced by an integral. The definition is thus, for "test functions" h,

$$\int \left.\frac{\delta H(f)}{\delta f}\right|_{f=F(\mathbf{x})} h(\mathbf{x})\, \mathrm{d}^n x = \lim_{\varepsilon \to 0} \frac{H\big(F(\mathbf{x}) + \varepsilon h(\mathbf{x})\big) - H\big(F(\mathbf{x})\big)}{\varepsilon}.$$

physically relevant situations. This would be more or less equivalent to the problem of whether the super wave functions, which should be elements of an appropriate Hilbert space, have a well defined unitary evolution.

This requires more elaboration. A Hilbert space of super wave functions would need a scalar product, which would be a generalisation of the known scalar product of the N-particle Hilbert space, viz.,

$$\langle \psi | \varphi \rangle = \int_{\mathbb{R}^{3N}} \psi^*(\mathbf{q}_1, \ldots, \mathbf{q}_N) \varphi(\mathbf{q}_1, \ldots, \mathbf{q}_N) \, d^{3N} q \, ,$$

to infinite dimensions. Recall that the meaning of $|\psi(\mathbf{q}_1, \ldots, \mathbf{q}_N)|^2 \, d^{3N}q$ is the probability distribution of the positions of the N particles. What expression could serve to generalise that in the infinite-dimensional space of field configurations? In the theory of measures and integration, we learn about the volume measure known as Lebesgue measure, but we should not be too surprised to find that no such infinite-dimensional Lebesgue measure exists. How could it? If we think about the volume of a cuboid as length times width times height and try to generalise that to infinitely many coordinate axes, we could only get zero, one, or infinity. In short, no such (translation and rotation invariant) volume measure exists on the function space $\mathcal{D}$.

But measure and integration theory can nevertheless be adapted to function spaces. It may surprise the reader to learn that the theory was also founded by Albert Einstein, through his work on Brownian motion, which eventually led to the acceptance of atomism in modern physics. We can't go into much detail here, but recall that a Brownian particle follows an erratic path—in the mathematical description, it is actually continuous, but nowhere differentiable. On such erratic trajectories, we can indeed construct a measure, which is in fact an infinite-dimensional Gaussian measure on the continuous functions on $\mathbb{R}$. This is called the Wiener measure μ_W, named after Norbert Wiener (1894–1964), who did pioneering work on this problem.

To flesh out these remarks a little, let us write down as an example the Wiener measure μ_W on a so-called cylinder set. This is a set of paths specified by finitely many time points $t_1, \ldots, t_n$ at which the paths $B : \mathbb{R} \to \mathbb{R}$ take values within specified sets, for example, within the infinitesimal intervals $dx_1, \ldots, dx_n$:

$$\mu_W\big(\{B(t_1) \in dx_1, B(t_2) \in dx_2, \ldots, B(t_n) \in dx_n\}\big)$$

$$= \frac{\exp\left(-\dfrac{x_1^2}{2Dt_1}\right)}{\sqrt{2\pi D t_1}} \frac{\exp\left[-\dfrac{(x_2 - x_1)^2}{2D(t_2 - t_1)}\right]}{\sqrt{2\pi D(t_2 - t_1)}} \cdots \frac{\exp\left[-\dfrac{(x_n - x_{n-1})^2}{2D(t_n - t_{n-1})}\right]}{\sqrt{2\pi D(t_n - t_{n-1})}} \, dx_1 \ldots dx_n \, .$$

This is an n-dimensional Gaussian measure (actually the probability for paths being in that cylinder set) with "diffusion constant" $D > 0$. If we decrease the distances between the base points t_i to zero, while increasing their number accordingly to approximate the continuum, we can consider "tubes" of possible paths in function space, and the Wiener measure determines the "content" or "volume" of these tubes.

With a little practice on Gaussian integrations, it is straightforward to compute the expected value:

$$\mathbb{E}\left[(B(t) - B(s))^2\right] = \int_{\mathbb{R}} (x_2 - x_1)^2 \frac{\exp\left[-\dfrac{(x_2 - x_1)^2}{2D(t_2 - t_1)}\right]}{\sqrt{2\pi D(t_2 - t_1)}} dx_2 = D \cdot (t - s).$$

Heuristically, this suggests that $B(t) - B(s)$ behaves as $\sqrt{t - s}$, which then implies that

$$\frac{B(t) - B(s)}{t - s} \sim \frac{1}{\sqrt{t - s}},$$

and hence that the quotient goes to infinity when $t \to s$. However, that just means that the Brownian paths are *nowhere differentiable*, as already pointed out.[10] Denoting the set of continuous functions by $\mathscr{C}$, we can use the Wiener measure to define the Hilbert space $L^2(\mathscr{C}, d\mu_W)$ of square-integrable super wave functions on "Brownian paths", i.e., on fields on $\mathbb{R}$.

By analogy, we can attempt first to construct a Gaussian measure on the physical field configurations, so that we can then define the absolute square of the super wave function as the probability density of the field configurations. The aim is to get a Hilbert space of super wave functions that are square integrable with respect to the Gaussian measure μ (which needs to be constructed) on the field space $\mathscr{D}$. In short, we try to construct the Hilbert space $L^2(\mathscr{D}, d\mu)$.

One insight which grew out of this program is that the typical field configurations of fields on $\mathbb{R}^3$ are not functions but rather distributions. This means that the relevant field space $\mathscr{D}$ which we introduced so nonchalantly above is rather "ugly" and mathematically abstract. This should not be too surprising when we recall the non-differentiability of Brownian paths, which can be seen as fields on $\mathbb{R}$. If we recall the construction of the Lebesgue measure, we may also recall that the "nice" points on the real axis, namely the rational numbers, form a null set of the Lebesgue measure. The "support" of the measure consists of the irrational numbers, and in fact only the transcendental numbers among those.

What do we learn from that? The typical field configurations are really very wild objects and the hope of seeing particles as "bumps" within such wild fields becomes rather vain.

Thus the question remains: How can we get to the famous photons, or light particles, in such quantum theories of fields? That comes about because, at least in "simple" models, the Hilbert space $L^2(\mathscr{D}, d\mu)$ description in terms of field configurations can be mapped to a description involving particle numbers, i.e., a Fock space description. In the next section, we shall study in more detail the fermionic Fock space, which is a direct sum of antisymmetric (fermionic) wave

---

[10]In fact, the trajectories are Hölder-continuous with exponent 1/2, i.e., $|B(t) - B(s)| \sim |t - s|^{1/2}$.

functions. In contrast, the Fock space which describes bosons consists of a direct sum of symmetric wave functions. Each summand represents a certain particle number. The map is defined on basis elements of the Hilbert space $L^2(\mathscr{D}, d\mu)$. These can be expressed in terms of Hermite polynomials (indexed by multi indices) which, in the simplest one-dimensional version, are the eigenfunctions of the harmonic oscillator studied in elementary courses on quantum mechanics. The multi indices $(n_1, \ldots, n_k)$, $k \in \mathbb{N}$, can be ordered by their sum $N = \sum_{i=1}^{k} n_i$, and the Hermite polynomials with multi-index sum N are mapped to the basis elements of the N-particle sector of Fock space. The reason why the multi-indexed Hermite polynomials form a basis in the infinite-dimensional Hilbert space lies with the structure of the functional Schrödinger equation for super wave functions, which is a linear (albeit infinite-dimensional) wave equation and which, under Fourier transform, can be viewed as an equation for infinitely many harmonic oscillators.

We must leave things like this, firstly because a proper mathematical formulation would vastly increase the length of this book and secondly because such a formulation would still only be an unfinished programme. The latter goes by the name of "constructive field theory" and falls within Dirac's second class of difficulties.

Different approaches or ansatzes can thus be adopted to formulate quantum field theory, but the infinities which are so bothersome and which we would like to sweep under the carpet will always resurface in one way or another—at least they have done up until now. At the end of the day we must realize that the problem is not just a quest for the right mathematical language, but that relativistic quantum theory will require new physical insights. The infinities which appear abundantly in the programme are not merely mathematical problems that we can try to solve by new techniques. They are clear signs of fundamental physical problems which will lead to a distortion of the notion of physical theory if we keep trying to push them to one side: non-convergent perturbation expansions and unspeakable limits of renormalization procedures become the substitute for perhaps two or three fundamental equations which would take up three lines on a piece of paper, as for example in Newtonian mechanics. To get from such fundamental equations to observable phenomena, experience teaches us to expect hard, perhaps extremely hard, analysis to be involved, but as physicists, we would have no reason to shy away from such efforts if the physical theory lay clearly before our eyes.

11.3 Fermionic Fock Space

This is a further mathematical insertion which will not really help us come to grips with the decisive questions. It is done for two purposes. First, we wish to understand some of the common notions of relativistic quantum physics—Fock space, second quantisation, and creation and annihilation operators—at least well enough to demystify them. Second, we would like to make it clear that the description of pair creation and annihilation in terms of the Dirac sea is indeed equivalent to the more common language of particles and antiparticles.

In relativistic physics, particles can be "created" and "annihilated". To describe this mathematically, we use a Hilbert space with variable particle number. This is called a *Fock space*. Such a description is often referred to as *second quantisation*, a foolish terminology that does not really mean more than what we just said: particles can be created and annihilated.

When we construct the Fock space, we wish to take into account at the outset the fact that the particles, electrons, are fermions, which are described by antisymmetric wave functions (see Sect. 4.4). To do so we first introduce the *antisymmetric product* of wave functions. Let $\mathcal{H} = L^2(\mathbb{R}^3, \mathbb{C}^k)$ denote the one-particle Hilbert space. For two vectors $\varphi_1, \varphi_2 \in \mathcal{H}$, the antisymmetric product is defined as

$$\varphi_1 \wedge \varphi_2 := \frac{1}{\sqrt{2}} (\varphi_1 \otimes \varphi_2 - \varphi_2 \otimes \varphi_1) .$$

On the right-hand side, we have the tensor product of the two vectors (the spin degrees of freedom are multiplied). If we focus on the position degrees of freedom, we can read this as a pointwise product, that is,

$$\varphi_1 \wedge \varphi_2(x, y) = \frac{1}{\sqrt{2}} \big[\varphi_1(x)\varphi_2(y) - \varphi_2(x)\varphi_1(y) \big] .$$

For N particles, the general expression is

$$\varphi_1 \wedge \varphi_2 \wedge \ldots \wedge \varphi_N = \frac{1}{\sqrt{N!}} \sum_{\sigma \in S_N} (-1)^\sigma \varphi_{\sigma(1)} \otimes \varphi_{\sigma(2)} \otimes \ldots \otimes \varphi_{\sigma(N)} . \qquad (11.7)$$

Here, we sum over all possible permutations where $(-1)^\sigma$ is positive or negative depending on whether the permutation consists of even or odd number of transpositions (commutation of two indices). This may be familiar from the definition of the determinant. However, it is sufficient to remember that commuting two indices yields a minus sign:

$$\varphi_1 \wedge \ldots \wedge \varphi_i \wedge \ldots \wedge \varphi_j \wedge \ldots \wedge \varphi_N = -\varphi_1 \wedge \ldots \wedge \varphi_j \wedge \ldots \wedge \varphi_i \wedge \ldots \wedge \varphi_N ,$$

and if two vectors are equal (or more generally if $\varphi_1, \ldots, \varphi_N$ are linearly dependent), the antisymmetric product is zero. If $\{\varphi_1, \varphi_2, \varphi_3, \ldots\}$ is an arbitrary basis of the one-particle Hilbert space $\mathcal{H}$, then the N-particle Hilbert space for fermions is spanned by the N-fold antisymmetric products:

$$\bigwedge^N \mathcal{H} := \mathrm{span}\big\{\varphi_{i_1} \wedge \varphi_{i_2} \wedge \ldots \wedge \varphi_{i_N} \mid i_1 < i_2 < \ldots < i_N \big\}. \qquad (11.8)$$

But we should always bear in mind that a typical N-particle state is not a product state but a linear combination (a superposition) of product states.

We now define the *fermionic Fock space* by forming the direct sum over all N-particle Hilbert spaces with $0 \leq N < \infty$:

$$\mathscr{F} := \bigoplus_{N=0}^{\infty} \bigwedge^N \mathscr{H} . \tag{11.9}$$

To understand this construction, the reader is advised to think about a typical element of this space. In position space and at a given time t, the wave function is defined on the disjoint union of configuration spaces $\bigsqcup_{N=0}^{\infty} \mathbb{R}^{3N}$, i.e.,

$$\Psi(x_1^1; x_1^2, x_2^2; x_1^3, x_2^3, x_3^3; \ldots, t) = (c, \phi_1(x_1^1, t), \phi_2(x_1^2, x_2^2, t), \phi_3(x_1^3, x_2^3, x_3^3, t), \ldots) .$$

It is a superposition of states with different numbers of particles. The "zero particle Hilbert space" is isomorphic to the field of complex numbers $\mathbb{C}$, in which 1 is called the *vacuum state*. This should simply be accepted as a useful mathematical convention.

Remark 11.2 Note that the Fock space states are defined at a common time since the notion of configuration space $\mathbb{R}^{3N}$ presupposes an absolute simultaneity (otherwise we would need to introduce configurations of spacetime, something we shall address briefly in Sect. 11.4). Therefore the construction is manifestly non-relativistic. In scattering theory, however, where we compute transition probabilities between $t = -\infty$ and $t = +\infty$, this is unimportant because they involve only asymptotically freely evolving states which lead easily to relativistically invariant expressions.

Since the Fock space contains states with different numbers of particles, it is helpful in describing the effects of particle creation and annihilation in an effective way. To this end, it is useful to introduce maps which transform an N-particle state into a state with $N + 1$ or $N - 1$ particles. These are the so-called *creation* and *annihilation* *operators*. For $\chi \in \mathscr{H}$, we define

$$a^*(\chi)\varphi_{i_1} \wedge \ldots \wedge \varphi_{i_N} := \chi \wedge \varphi_{i_1} \ldots \wedge \varphi_{i_N} , \tag{11.10}$$

$$a(\chi)\varphi_{i_1} \wedge \ldots \wedge \varphi_{i_N} := \sum_{k=1}^{N} (-1)^{k+1} \langle \chi, \varphi_{i_k} \rangle \varphi_{i_1} \wedge \ldots \wedge \cancel{\varphi_{i_k}} \wedge \ldots \wedge \varphi_{i_N} . \tag{11.11}$$

For practice, the reader should check the anticommutation relation

$$\{a(\chi), a^*(\phi)\} = \langle \chi, \phi \rangle .$$

The common manner of speaking is to say that the operator $a^*(\chi)$ "creates" a particle in state χ and the operator $a(\chi)$ annihilates the indicated particle. Of course, it is easy to understand why we might put things this way, but this manner of speaking should be taken with a grain of salt. Clearly, an operator does not create a particle; it is a mathematical notation.

With the help of creation and annihilation operators, we can lift linear maps from $\mathscr{H}$ to Fock space. For example, the operator $\hat{A} = \sum_{i,j} \alpha_{ij} |\varphi_j\rangle\langle\varphi_i|$ on $\mathscr{H}$ becomes the operator $\hat{\hat{A}} = \sum_{i,j} \alpha_{ij}\, a^*(\varphi_j) a(\varphi_i)$ on $\mathscr{F}$. This is what is referred to as "second quantisation" of the operator, presumably because we put a second hat on A. This may seem at first sight a bit out of proportion, but it is at the end of the day a useful mathematical formalism. There is no reason to feel intimidated by it.

11.3.1 Particles and Antiparticles

We now wish to discuss a complication when using the Fock space description of the Dirac theory, i.e., quantum electrodynamics (QED). We have already seen that the Dirac Hamiltonian (11.1) has a negative energy spectrum which is not bounded from below. Coupling the Dirac equation to an electromagnetic field would lead to the *radiation catastrophe*, because a single electron could radiate an unlimited amount of energy by sinking into ever more negative energies. We talked about that in connection with the Dirac sea picture. We shall pick up on that discussion in connection with the Fock space construction.

The standard *ad hoc* solution within the Fock space construction consists in transforming the negative energy states into positive energy states with opposite charge. These *antiparticles* are called the *positrons*. This is done by introducing what is called *charge conjugation*, i.e., an anti-unitary map $\mathscr{C}$ for which the following holds: if ψ is a solution of the Dirac equation with negative energy then $\mathscr{C}\psi$ is a solution of the Dirac equation with positive charge and positive energy. We decompose the one-particle Hilbert space into subspaces of positive and negative energy with respect to the free Dirac hamiltonian D_0:

$$\mathscr{H} = \mathscr{H}_+ \oplus \mathscr{H}_- . \tag{11.12}$$

In the particle/antiparticle picture, we then have the electron states in $\mathscr{H}_+$ and the positron states in $\mathscr{C}\mathscr{H}_-$. The Fock space of the particle part of QED is now defined as

$$\mathscr{F} = \bigoplus_{m+n=N,\, N=0,\dots,\infty} \bigwedge^m \mathscr{H}_+ \otimes \bigwedge^n \mathscr{C}\mathscr{H}_- . \tag{11.13}$$

The *vacuum* is the "zero-particle" state $\Omega := 1 \otimes 1 \in \mathbb{C} \otimes \mathbb{C}$ (for $m = n = 0$).

Let $\{e_1, e_2, e_3, \dots\}$ be a basis of $\mathscr{H}_+$ and $\{e_0, e_{-1}, e_{-2}, e_{-3}, \dots\}$ a basis of $\mathscr{H}_-$. Think, for example, of energy eigenfunctions, ordered according to their eigenvalues. A typical Fock space state is now a linear combination of product wave functions of the form

$$e_{i_1} \wedge e_{i_2} \wedge \ldots \wedge e_{i_m} \wedge \mathscr{C} e_{j_1} \wedge \mathscr{C} e_{j_2} \wedge \ldots \wedge \mathscr{C} e_{j_n}, \tag{11.14}$$

where $0 < i_1 < i_2 < \ldots < i_m$ and $0 \geq j_1 > j_2 > \ldots > j_n$. Here, the number of particles $N = m + n$ is variable (but the charge $c = n - m$ is a conserved quantity and hence constant). In terms of the creation operators,

$$a^*(e_k) e_{i_1} \wedge e_{i_2} \wedge \ldots \wedge e_{i_m} := e_k \wedge e_{i_1} \wedge e_{i_2} \wedge \ldots \wedge e_{i_m},$$
$$b^*(e_{-k}) \mathscr{C} e_{j_1} \wedge \mathscr{C} e_{j_2} \wedge \ldots \wedge \mathscr{C} e_{j_n} := (-1)^m \mathscr{C} e_{-k} \wedge \mathscr{C} e_{j_1} \wedge \ldots \wedge \mathscr{C} e_{j_n},$$

where $k > 0$ and m denotes the number of electrons, we can write (11.14) in the form

$$a^*(e_{i_1}) \cdots a^*(e_{i_m}) b^*(e_{j_1}) \cdots b^*(e_{j_n}) \, \Omega. \tag{11.15}$$

In the manner of speaking introduced above, $a^*(\cdot)$ generates an electron state and $b^*(\cdot)$ a positron state, and we obtain the basis vectors of the fermionic Fock space by successively applying these creation operators to the vacuum state Ω.

11.3.2 Fock Space as the Dirac Sea

Finally, we return once again to the Dirac sea in its mathematically rigorous form to construct the Fock space from it. The physical picture is now different from the one above. Instead of a variable number of electrons and positrons, we now have an infinite number of electrons which occupy all negative states apart from a finite number. It may be of some comfort to note that, in this approach, particles do not get created from nothing and do not vanish into nothing. In this theory, these particles are always there, albeit not visible to us.[11] Moreover, this approach yields a clearer understanding of the "second class" difficulty in constructing QED on a Fock space. We shall say a bit more about that at the end of this section.

We write the ground state of the Dirac sea—the state in which all negative states are occupied and no electrons are visible—as an infinite antisymmetric product:

$$\tilde{\Omega} = e_0 \wedge e_{-1} \wedge e_{-2} \wedge e_{-3} \wedge \ldots. \tag{11.16}$$

Anyone with mathematical sensibilities will be concerned as to whether such an infinite expression is well defined. However, this is not problematic because the

---

[11]D.-A. Deckert, M. Esfeld, and A. Oldofredi, A persistent particle ontology for QFT in terms of the Dirac sea. In the British Journal for the Philosophy of Science, online version: arXiv:1608.06141, 2016.

antisymmetric product of countably many vectors can be identified (up to a constant, which means projectively) with the subspace spanned by these vectors. In this sense, the infinite product corresponds to the subspace $\mathscr{H}_-$ of the negative energy states and, up to a constant, (11.16) is independent of the choice of basis.[12]

Furthermore, creation and annihilation operators can be introduced for the Dirac sea. For $e_k \in \mathscr{H}_+$, let $\tilde{a}^*(e_k)$ add an electron with positive energy state e_k to the sea, i.e.,

$$\tilde{a}^*(e_k)\, e_0 \wedge e_{-1} \wedge e_{-2} \wedge \ldots := e_k \wedge e_0 \wedge e_{-1} \wedge e_{-2} \wedge \ldots .$$

For $e_{-k} \in \mathscr{H}_-$, let $\tilde{b}(e_{-k})$ erase the particle in the negative energy state e_{-k}, i.e.,

$$\tilde{b}(e_{-k})\, e_0 \wedge e_{-1} \wedge e_{-2} \wedge \ldots := (-1)^k e_0 \wedge e_{-1} \wedge \ldots \wedge e_{\cancel{-k}} \wedge e_{-(k+1)} \wedge \ldots , \quad (11.17)$$

where the expression is set to zero if the left-hand side does not contain the state e_{-k}. The possible states of the Dirac sea are now linear combinations of states of the form

$$\tilde{a}^*(e_{i_1}) \cdots \tilde{a}^*(e_{i_m}) \tilde{b}(e_{j_1}) \cdots \tilde{b}(e_{j_n})\, \tilde{\Omega} , \quad (11.18)$$

where $0 < i_1 < \ldots < i_m$ and $0 \geq j_1 > \ldots > j_n$. This should be compared with (11.15). Such a state contains m electrons of positive energy and n "holes", which are unoccupied states of negative energy. The number of visible particles (including the holes) is thus $N = m + n$ and the net charge relative to the ground state $\tilde{\Omega}$ is $c = n - m$.

By defining the linear map

$$\begin{aligned}
F : \tilde{a}^*(e_{i_1}) \cdots \tilde{a}^*(e_{i_m}) \tilde{b}(e_{j_1}) \cdots \tilde{b}(e_{j_n})\, \tilde{\Omega} \\
\longmapsto a^*(e_{i_1}) \cdots a^*(e_{i_m}) b^*(e_{j_1}) \cdots b^*(e_{j_n}) \Omega ,
\end{aligned} \quad (11.19)$$

we see at once that the Hilbert space of the Dirac sea states is isomorphic to the fermionic Fock space. Specifically, the ground state $\tilde{\Omega}$ corresponds to the vacuum Ω. The occupied states of positive energy correspond to electrons and the holes in the Dirac sea become positrons with positive energy and charge in the particle/antiparticle language.

We note also that, in this isomorphism, the annihilation operator $\tilde{b}(\cdot)$ becomes a positron creation operator $b^*(\cdot)$. The physical intuition behind this is now clear. From a mathematical point of view, however, a consistency check is important, for the following reason. A creation operator is normally *linear* in its argument, which means, for example, that $a^*(\lambda\varphi + \psi) = \lambda a^*(\varphi) + a^*(\psi)$ holds for arbitrary $\lambda \in \mathbb{C}$, and analogously for $\tilde{a}^*$. An annihilation operator is normally anti-linear in

[12]D. Lazarovici, *Time evolution in the external field problem of quantum electrodynamics*. Thesis, LMU München, 2011. Online version: arXiv:1310.1778.

its argument, i.e., $\tilde{b}(\lambda\varphi + \psi) = \lambda^*\tilde{b}(\varphi) + \tilde{b}(\psi)$, where λ^* denotes the complex conjugate. The positron creation operator $b^*(\cdot)$ is also anti-linear, however, because it involves the anti-linear charge conjugation $\mathscr{C}$ [see (11.17)]. What does this tell us? Perhaps just that the "creation" of a positron really is the creation of a hole left when an electron has been lifted from the Dirac sea.

Remark 11.3 (Renormalization of Infinite Pair Creation) As already mentioned, a magnetic field typically lifts infinitely many particles from the Dirac sea, i.e., the Dirac sea comes with infinite pair creation, since charged particles like electrons interact via electromagnetic radiation and hence magnetic fields are always present. In the Fock space description which is commonly used nowadays, there is only the vacuum from which particles are created, and it would be easy to think that such Dirac sea infinities have been overcome. That is why this section is of special importance: since the two descriptions are mathematically equivalent, it is clear that some infinities must appear in the Fock space description as well. In the latter, this is hidden in the assertion that the quantum mechanical evolution of interacting fermions cannot be constructed on one and the same Fock space

The remedy for this can be understood very simply in the Dirac sea picture. As we saw above, a magnetic field "rotates" the negative energy states to positive energy states, so infinitely many particles with positive energy appear. But then why not just take the "rotated" Dirac sea as the new Dirac sea, so that at best finitely many particles are "above the sea"? In Fock space language, that would mean that the vacuum has to be readjusted or redefined. Roughly speaking, the vacuum and with it the whole Fock space changes with the changing magnetic field (for a mathematical formulation see, for example, the references in footnote 7).

11.4 Multi-Time Wave Function

We shall come back to the question of first class difficulties in the context of relativistic quantum mechanics in Chap. 12, but for the moment there is one more technical question we would like to discuss. We saw in the earlier chapters that the revolution in quantum mechanics lies in the wave function leading to entanglement and nonlocality. In non-relativistic physics, the wave function is defined on the configuration space of an N-particle system. But this configuration space is a blatantly non-relativistic construct because it contains all possible configurations of the system at a *common* time and hence presupposes an absolute simultaneity which is not part of relativistic physics. The question which we wish to approach now no longer concerns the creation or annihilation of particles, but rather how we can come to a serious relativistic description of the wave function for a given particle number N.

We can approach the problem in more general terms. A relativistic theory (we shall always have in mind special relativity here) must be invariant under Lorentz transformations $x = (t, \mathbf{x}) \mapsto \Lambda x$, which describe a change of reference frame in the four-dimensional Minkowski spacetime. Then a one-particle wave function $\varphi(t, \mathbf{x}) = \varphi(x)$ in relativistic spacetime transforms under Lorentz transformation according to

$$\varphi(x) \xrightarrow{\Lambda} S[\Lambda]\varphi(\Lambda^{-1}x), \tag{11.20}$$

where Λ^{-1} acts as a 4×4 matrix and $S[\Lambda]$ denotes the representation of the Lorentz transformation acting on the spinor components.

For an N-particle system, we can express the wave function at a given time t as

$$\psi(t, \mathbf{x}_1, \mathbf{x}_2, \ldots, \mathbf{x}_N) = \psi(t, \mathbf{x}_1, t, \mathbf{x}_2, \ldots, t, \mathbf{x}_N). \tag{11.21}$$

Then note that the N particle coordinates transform under Lorentz transformation according to

$$(t, \mathbf{x}_1, \ldots, t, \mathbf{x}_N) \xrightarrow{\Lambda} \left(\Lambda^{-1}(t, \mathbf{x}_1), \ldots, \Lambda^{-1}(t, \mathbf{x}_N)\right) = (\tilde{t}_1, \tilde{\mathbf{x}}_1, \ldots, \tilde{t}_N, \tilde{\mathbf{x}}_N), \tag{11.22}$$

where the times $\tilde{t}_1, \ldots, \tilde{t}_N$ are now generally *different* (if $\mathbf{x}_1, \ldots, \mathbf{x}_N$ are different). In the new Minkowski coordinates, the N-particle wave function ψ would have to assume the form

$$\psi(t, x_1, t, x_2, \ldots, t, x_N) \xrightarrow{\Lambda} \underbrace{S[\Lambda] \otimes \ldots \otimes S[\Lambda]}_{N \text{ times}} \psi\left(\Lambda^{-1}(t, x_1), \ldots, \Lambda^{-1}(t, x_N)\right)$$

$$= S[\Lambda] \otimes \ldots \otimes S[\Lambda]\psi\left(\tilde{t}_1, \tilde{\mathbf{x}}_1, \ldots, \tilde{t}_N, \tilde{\mathbf{x}}_N\right), \tag{11.23}$$

which is no longer an N-particle wave function at a common time. The obvious solution, which also goes back to Dirac, is now simply to consider the wave function (11.21) as a special case of the more general *multi-time wave function*

$$\psi : \Gamma \subseteq \underbrace{\mathbb{R}^4 \times \ldots \times \mathbb{R}^4}_{N-\text{times}} \to \underbrace{\mathbb{C}^4 \otimes \ldots \otimes \mathbb{C}^4}_{N-\text{times}}, \quad (x_1, \ldots, x_N) \mapsto \psi(x_1, \ldots, x_N),$$

in which each position coordinate is supplemented with its own time coordinate

$$\psi(x_1, \ldots, x_N) = \psi(t_1, \mathbf{x}_1, \ldots, t_N, \mathbf{x}_N).$$

In view of the single-time version (11.21), the natural domain of definition is not the whole of $\mathbb{R}^{4N}$, but the subset of *spacelike configurations*

$$\Gamma := \left\{ (x_1, \ldots, x_N) \in \mathbb{R}^{4N} \mid \forall i \neq j : (x_i - x_j)^2 < 0 \right\}, \tag{11.24}$$

where $(x_i - x_j)^2 = (t_i - t_j)^2 - (\mathbf{x}_i - \mathbf{x}_j)^2$ denotes the scalar product with respect to the Minkowski metric. Every $(x_1, \ldots, x_N) \in \Gamma$ lies in general in a curved spacelike hypersurface $\Sigma \subset \mathcal{M}$, where $\mathcal{M}$ is Minkowski spacetime.

On the other hand, we can start with $(x_1, \ldots, x_N) \in \Sigma^N$ and introduce the multi-time wave function as $\psi_\Sigma(x_1, \ldots, x_N)$ by thinking in the following way. The non-relativistic wave function is indexed by the time parameter t—corresponding to a foliation of Newtonian spacetime into absolute simultaneity surfaces—while the relativistic multi-time wave function is defined on general spacelike hypersurfaces.

To define appropriate Hilbert spaces, we introduce first the N-particle "tensor current"

$$j_\psi^{\mu_1 \cdots \mu_N}(x_1, \ldots, x_N) := \overline{\psi}(x_1, \ldots, x_N) \gamma_1^{\mu_1} \cdots \gamma_N^{\mu_N} \psi(x_1, \ldots, x_N), \qquad (11.25)$$

with $\overline{\psi} = \psi^* \gamma_1^0 \ldots \gamma^0$, where the matrices $\gamma_i^{\mu_i}$ act on the spinor component of the i th particle. We can integrate the current through a hypersurface Σ as follows:

$$\int_\Sigma d\sigma_1(x_1) \ldots \int_\Sigma d\sigma_N(x_N)\, n_{\mu_1}(x_1) \ldots n_{\mu_N}(x_N) j_\psi^{\mu_1 \cdots \mu_N}(x_1, \ldots, x_N),$$
$$(11.26)$$

where $d\sigma$ denotes the surface element and n_μ is the vector field normal to Σ, which is future directed and orthogonal to the hypersurface at every point. Thus, for every configuration $(x_1, \ldots, x_N) \in \Sigma^N$, the contraction

$$\rho_\Sigma(x_1, \ldots, x_N) := n_{\mu_1}(x_1) \ldots n_{\mu_N}(x_N) j_\psi^{\mu_1 \cdots \mu_N}(x_1, \ldots, x_N) \qquad (11.27)$$

defines a *crossing density*, integrated over the hypersurface in (11.26) (see Fig. 11.1).

It should now be checked that, if Σ is a hypersurface with constant t coordinates in a given frame of reference, i.e., $\Sigma = \{(t, \mathbf{x}) : \mathbf{x} \in \mathbb{R}^3\}$, then (11.26) yields the standard $|\psi|^2$ norm

$$\int d^3x_1 \ldots \int d^3x_N\, \psi^*(x_1, \ldots, x_N) \psi(x_1, \ldots, x_N). \qquad (11.28)$$

This can be taken as the motivation for normalizing the wave function in the general case by setting the quantity in (11.26) equal to 1 and interpreting the quantity in (11.27) as a probability density, by analogy with Born's rule, although this interpretation is not unproblematic, as we shall see later (Remark 12.1).

Moreover, by analogy with (11.28), we would like to read (11.26) as a scalar product. To do so, as a generalisation of (11.25), we introduce the tensor density

$$j^{\mu_1 \cdots \mu_N}[\phi, \psi](x_1, \ldots, x_N) := \overline{\phi}(x_1, \ldots, x_N) \gamma_1^{\mu_1} \cdots \gamma_N^{\mu_N} \psi(x_1, \ldots, x_N).$$
$$(11.29)$$

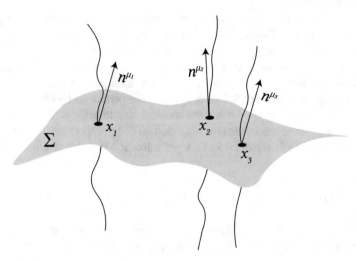

Fig. 11.1 The tensor current defines a crossing density which gets integrated over the hypersurface. Three flux lines are drawn as an example

This is a quadratic form in the two wave functions ϕ, ψ, such that

$$j_\psi^{\mu_1 \cdots \mu_N} = j^{\mu_1 \cdots \mu_N}[\psi, \psi].$$

This allows us to define the Hilbert spaces

$$\mathscr{H}_\Sigma := L^2(\Sigma^N, \langle \cdot, \cdot \rangle_\Sigma), \tag{11.30}$$

with the Lorentz invariant scalar product

$$\langle \phi, \psi \rangle_\Sigma := \int_\Sigma \mathrm{d}\sigma_1(x_1) \ldots \int_\Sigma \mathrm{d}\sigma_N(x_N)\, n_{\mu_1}(x_1) \ldots n_{\mu_N}(x_N)\, j^{\mu_1 \cdots \mu_N}[\phi, \psi](x_1, \ldots, x_N).$$

Hence the wave functions are normalized according to

$$\|\psi\|_\Sigma := \sqrt{\langle \psi, \psi \rangle_\Sigma} = 1.$$

11.4.1 Time Evolution

It is appropriate now to talk about the dynamics of the multi-time wave function. We would like to have a relativistic wave equation which defines a unitary evolution between spacelike hypersurfaces, i.e., a family of unitary operators $U(\Sigma, \Sigma')$:

$\mathscr{H}_{\Sigma'} \rightarrow \mathscr{H}_{\Sigma}$ such that, for arbitrarily chosen spacelike hypersurfaces $\Sigma, \Sigma', \Sigma'' \subset \mathscr{M}$, we have:

i) $U(\Sigma, \Sigma) = \mathbf{1}$,
ii) $U(\Sigma, \Sigma')U(\Sigma', \Sigma'') = U(\Sigma, \Sigma'')$,
iii) $\|U(\Sigma, \Sigma')\psi\|_\Sigma = \|\psi\|_{\Sigma'}$.

Here, unitarity is tantamount to conservation of the N-particle current (11.25). In relativistic notation, we should thus obtain N continuity equations

$$\partial_{\mu_k} j^{\mu_1 \cdots \mu_k \cdots \mu_N} = 0, \quad k = 1, \ldots, N, \tag{11.31}$$

from which we must be able to derive the conservation of the total measure (11.26).

However, we are forced to admit that we do not know what the appropriate law is—an empirically adequate and mathematically consistent equation for a many-particle wave function has not yet been found. What we can do at least is to write down the *free* evolution, i.e., without interaction. It is given by a system of N Dirac equations, one for each time coordinate:

$$i\frac{\hbar}{c}\frac{\partial}{\partial t_1}\psi = H_1\psi,$$

$$i\frac{\hbar}{c}\frac{\partial}{\partial t_2}\psi = H_2\psi,$$

$$\vdots \tag{11.32}$$

$$i\frac{\hbar}{c}\frac{\partial}{\partial t_N}\psi = H_N\psi,$$

where

$$H_k = D^0 = -\sum_{l=1}^{3} \alpha_l^k i\hbar \frac{\partial}{\partial_{k,l}} + mc\beta \tag{11.33}$$

is the free Dirac Hamiltonian (without fields) [see (11.1)] acting on the coordinates of the k'th particle. In the free theory, we can easily check all desired properties—for example, the invariance of the measure—but this is at best of academic interest since we obviously live in a world with interactions, e.g., electromagnetic interactions.

The question is thus once again: How can we describe interactions in a consistent manner in the relativistic spacetime? There have been attempts to answer this question, but we cannot discuss them in detail here. In any case, a satisfactory answer is still lacking. The canonical ansatz of quantum field theory consists in coupling the Dirac Hamiltonian to an electromagnetic field and second quantising that field to get photons which are created and annihilated, thereby providing an interaction between the electrons. Although the theory is not well defined, as it contains the ultraviolet divergencies we talked about earlier, these can be circumvented by renormalization tricks. We shall leave things like this and simply admit that we do not know what the correct theory is.

Further Food for Thought 12

[The usual quantum] paradoxes are simply disposed of by the 1952 theory of Bohm, leaving as *the* question, the question of Lorentz invariance. So one of my missions in life is to get people to see that if they want to talk about the problems of quantum mechanics – the real problems of quantum mechanics – they must be talking about Lorentz invariance.

John S. Bell, Interview with Renée Weber[1]

We have omitted a lot in our reconstruction of the current state of relativistic quantum theory. For instance, the path integral formalism that is often hailed as a manifestly relativistic version of quantum field theory while, in fact, it is merely a technical reformulation that has just as many problems. Neither did we speak about string theory and other approaches to a quantum theory of gravity that are beyond the scope of this book. On the other hand, our critical assessment of quantum field theory may have created the impression that we are insisting on pointless mathematical rigor in the formulation of physical theories. We are not. Throughout this book, we have deliberately stayed well clear of purely academic questions like, for instance, the domain of self-adjointness of observable operators. Whoever sees in those a key to understanding quantum mechanics is certainly wrong. What we insist on are physical laws that make sense. Laws that do not merely describe asymptotic scattering states (after enough massaging) but tell us how the universe might actually work.

Here is Dirac once again, in the *Scientific American* article mentioned earlier:

It seems to be quite impossible to put this theory on a mathematically sound basis. At one time physical theory was all built on mathematics that was inherently sound. I do not say that physicists always use sound mathematics; they often use unsound steps in their calculations. But previously when they did so it was simply because of, one might say, laziness. They wanted to get results as quickly as possible without doing unnecessary work. It was always possible for the pure mathematician to come along and make the theory

---

[1]Quoted from: M. Bell and S. Gao (Eds.), *Quantum Nonlocality and Reality: 50 Years of Bell's Theorem*. Cambridge University Press, 2016, p. 369.

© Springer Nature Switzerland AG 2020
D. Dürr, D. Lazarovici, *Understanding Quantum Mechanics*,
https://doi.org/10.1007/978-3-030-40068-2_12

sound by bringing in further steps, and perhaps by introducing quite a lot of cumbersome notation and other things that are desirable from a mathematical point of view in order to get everything expressed rigorously but do not contribute to the physical ideas. The earlier mathematics could always be made sound in that way, but in the renormalization theory we have a theory that has defied all the attempts of the mathematician to make it sound. I am inclined to suspect that the renormalization theory is something that will not survive in the future, and that the remarkable agreement between its results and experiment should be looked on as a fluke.

The last few decades, which Dirac has not been able to witness, have provided continuous confirmation for both the remarkable empirical success of renormalized quantum field theory and Dirac's assessment that the theory evades a mathematically consistent formulation. The undeniable accomplishments of the Standard Model of particle physics, in particular, tell us that it must get a lot right and deserves to be studied in much more detail than would be possible in this book. The ultimate goal of physics, however, must be to embed the successful formalism into a fundamental theory that is mathematically sound and conceptually clear.

So let's assume that at some point in the near future, a new Einstein or Dirac comes along and finds the right equation for the relativistic multi-time wave function; an equation that describes interactions in relativistic spacetime while avoiding the problems of self-interactions and infinite pair creation. We still wouldn't be quite done, since we would still have to address the "first class problems", i.e., avoid the measurement problem and underpin relativistic wave mechanics with a clear ontology. In other words, we must understand how the three precise quantum theories described in this book (or at least one of them) can be generalized to the relativistic domain.

In this final chapter, we shall present possible approaches that are still incomplete but can provide readers with a more comprehensive perspective on the subject. As tentative or speculative as they may be, these approaches already show that we don't have to give up on a clear ontology and an objective description of nature, even when it comes to relativistic quantum physics. They also show, however, that there is still a lot of work to be done in order to arrive at theories that are as satisfying and well understood as those discussed in the case of non-relativistic quantum mechanics.

12.1   Many Worlds Interpretation of Relativistic Quantum Mechanics

Everettians like to say that the Many Worlds theory is the only version of quantum mechanics that also exists relativistically. That this cannot be true should be evident from the fact that the Many Worlds theory is defined in terms of the universal wave function and its unitary time evolution. Since we have no idea what relativistic law determines this time evolution, and hence the wave function "at finite times", there is, as of today, no relativistic Many Worlds theory. More to the point is the following statement: if we had a consistent, interacting, Lorentz invariant law for the relativistic wave function—analogous to Schrödinger's wave mechanics—we could apply a Many Worlds interpretation to it. Analysis of the relativistic wave equation should yield a similar picture to the one in the non-relativistic case: a branching (through decoherence) wave function of the universe in which we can try to describe a multitude of semi-classical worlds.

At this point, we should say, however, that there are different opinions about how exactly this splitting into worlds or histories should be understood. It is somewhat frustrating that the Many Worlds theory does not yield an unambiguous answer, but allows a variety of different interpretations, so to speak. Of these, we shall highlight two, as applied schematically to a simple example. We consider two distant systems, let's say a measurement setup ("the Φ-apparatus") with ready state Φ_0, and a second, spacelike separated system (the ψ-system) with wave function ψ_0. Suppose the measurement device carries out a "measurement" (on a third system, ignored in the following), thus evolving into a decoherent superposition of the possible measurement states Φ_1 and Φ_2. Before the Φ-apparatus and the distant ψ-system get to interact with each other, their common wave function (omitting the rest of the universe as well as normalization constants) evolves into

$$\Phi_0\psi_0 \longrightarrow (\Phi_1 + \Phi_2)\psi_0 = \Phi_1\psi_0 + \Phi_2\psi_0. \tag{12.1}$$

The question is now: Does the ψ-system exist once or twice?

From the $3N$-dimensional perspective, we would say that the right-hand side of (12.1) contains two decoherent branches of the wave function (Φ_1 and Φ_2 have macroscopically disjoint supports) which should correspond to two separate worlds. We thus have two copies of the ψ-system, one that exists in a world together with the Φ-apparatus pointing to "1", and one that exists in a world together with the Φ-apparatus pointing to "2". In this sense, the splitting into "worlds" would be a global process, affecting the entire universe at once. This view is hard to reconcile with relativity, though, for whether the measurement with the Φ-apparatus has already occurred or not will depend on the reference frame or, more generally, the hypersurface in which we look at the universal wave function.

We shall thus emphasize a different criterion and say this: in (12.1), the measurement apparatus has already split, but at this point, the distant system in the state ψ_0 can still interact with both branches Φ_1 and Φ_2. The ψ-system has thus not yet taken part in the branching, and it makes no sense to ask whether it exists in the Φ_1-world or in the Φ_2-world.[2]

This corresponds, at least roughly, to the view advocated in particular by David Wallace.[3] This view tries to conceive the Many Worlds theory in a more four-dimensional, i.e., spatiotemporal sense. In the spacetime region in which the measurement has occurred, there is the Φ-apparatus pointing to "1" *and* the Φ-

---

[2] More formally, we may say that the property of coexisting in a world is not transitive: Φ_1 and ψ_0 exist in the same world, Φ_2 and ψ_0 exist in the same world, yet Φ_1 and Φ_2 do not.

[3] See D. Wallace, *The Emergent Multiverse: Quantum Theory According to the Everett Interpretation*. Oxford University Press, 2012.

apparatus pointing to "2". In the spacelike separated region of the ψ-system, however, no branching has yet occurred, and the system (be it a second measurement device, a human "observer", or whatever) exists only once, in the state ψ_0. The branches of the apparatus wave function, however, will continue to interact with the environment, whether it be only air molecules or the photon field (the environment "measures" the pointer position, so to speak). Decoherence spreads, and after some time, it will come to affect the ψ-system. The total wave function is then something like $\Phi_1\psi_1 + \Phi_2\psi_2$, where ψ_1 and ψ_2 are entangled with different pointer states of the apparatus. According to Wallace, it is only now that we should say that the ψ-system has taken part in the branching, existing once in the state ψ_1 in a world together with Φ_1, and once in the state ψ_2 in a world with Φ_2.

The advantage of this view is that, with a relativistic wave equation, decoherence can spread at most with the speed of light, so that the branching of worlds, originating in local events, could spread at most with the speed of light. More basically, only to the extent that we can think about the Many Worlds theory in spatiotemporal terms does it make sense to speak of a seriously relativistic theory (otherwise it means nothing more than that the wave equation happens to have a Lorentz symmetry).

And yet, however we want to think about it, the notion of a "world" will always have a strikingly nonlocal character. Consider the EPR experiment with the two spin measurements occurring in spacelike separated regions A and B. We write the post-measurement wave function as

$$|\Psi\rangle = |\Uparrow\rangle_A |\Downarrow\rangle_B + |\Downarrow\rangle_A |\Uparrow\rangle_B , \qquad (12.2)$$

where $|\Uparrow\rangle_A$, $|\Downarrow\rangle_B$, etc., denote the macroscopic measurement devices in the corresponding spacetime regions that have registered "Spin UP" or "Spin DOWN". Now, we would have to say that the $|\Uparrow\rangle_A$ apparatus in region A and the $|\Downarrow\rangle_B$ apparatus in region B belong to one world, while the $|\Downarrow\rangle_A$ apparatus in A and $|\Uparrow\rangle_B$ in B belong to another. And this is a manifestly nonlocal effect. It has not originated from local interactions between the two devices, and nor can it be described as a local feature of spacetime regions A and B individually.

We conclude this difficult (and not completely settled) discussion with the following observation. To the extent that Many Worlds makes sense in the non-relativistic case, it could, in principle, be generalized to a relativistic quantum theory. The status of relativistic spacetime and localized objects in it remains somewhat obscure, or at least disputed. But from a purely technical point of view, a relativistic generalization is more straightforwardly obtained for Many Worlds than for Bohmian mechanics or GRW, since we need only a unitary evolution of the wave function and don't have to worry about the Lorentz invariant formulations of things like the Bohmian guiding equation or the GRW collapse law. Many Worlds could then be thought of as the price we have to pay for reconciling relativity and quantum physics.

Is this our only option? In fact, it is not. As we shall discuss in the following, relativistic generalizations of Bohmian mechanics and the GRW theory are possible, as well—but they too come with a price tag.

12.2 Relativistic Bohm–Dirac Theory

If we think back to Bohmian mechanics and GRW, the role of the wave function in these theories was first and foremost to determine and "synchronize" the motion and the appearance, respectively, of the ontological entities, i.e., the particles in Bohmian mechanics and the flashes in the GRWf theory. This synchronization, provided by the entangled wave function, was crucial to understanding the nonlocality that we discussed in detail in Chap. 10. At the same time, it makes the relativistic generalization of these theories highly non-trivial, since in relativistic spacetime, we no longer have hyperplanes of absolute simultaneity in which the motion of the Bohmian particles or the localization of the GRW flashes can be synchronized.

Consider first the Bohmian theory. In non-relativistic Bohmian mechanics, the particle configuration evolves along a vector field which is proportional to the quantum flux $j^\psi = (\mathbf{j}_1, \ldots, \mathbf{j}_N)$ determined by the wave function ψ. We can write this flux in a more relativistic notation as $j_\psi^{\mu_1 \ldots \mu_N} := j_1^{\mu_1} \otimes \ldots \otimes j_N^{\mu_N}$, $\mu_k \in 0, 1, 2, 3$, by setting the zero-component of the four-vectors to $j_k^0 = \rho = |\psi|^2$. The symbol $\otimes$ denotes the tensor product, but it can be thought of here as a simple multiplication of the coordinates for our present purposes. The Bohmian guiding equation (4.6) can now be written as:

$$\dot{Q}_k^{\mu_k}(t) = \frac{j_\psi^{0 \ldots \mu_k \ldots 0}}{j_\psi^{0 \ldots 0 \ldots 0}}(q_1, \ldots, q_N)\Bigg|_{q_i = Q_i(t)}, \tag{12.3}$$

with $Q_k^0(t) \equiv t$. The relativistic multi-time wave function still defines a conserved quantum flux in the form of the four-tensor (11.25). The classical velocity field, however, is evaluated at the positions of N particles *at the same time t*, and in relativistic spacetime, evaluation "at the same time t" makes no sense, since there is no absolute simultaneity. Therefore, in order to formulate a Bohmian theory, we need a distinguished family of hypersurfaces, i.e., a *foliation* of the four-dimensional Minkowski spacetime into three-dimensional spacelike submanifolds:

$$\mathfrak{F} := (\Sigma_t)_{t \in \mathbb{R}}, \quad \bigcup_{t \in \mathbb{R}} \Sigma_t \cong \mathcal{M}. \tag{12.4}$$

Given a spacelike hypersurface $\Sigma \in \mathfrak{F}$ and the worldline of the kth particle $X_k = X_k^{\mu_k}$, we denote by $X_k(\Sigma)$ the spacetime point at which it intersects the hypersurface Σ. The natural "relativistic generalization" of the Bohmian guiding equation thus takes the form

$$\frac{d}{d\tau_k} X_k^{\mu_k}(\tau_k) \propto j_\psi^{\mu_1 \cdots \mu_k \cdots \mu_N}(x_1, \ldots, x_N) \prod_{j \neq k} n_{\mu_j}(x_j) \Bigg|_{x_j = X_j(\Sigma)}. \qquad (12.5)$$

Here, τ_k denotes the proper time of particle k, and the proportionality factor on the right-hand side is chosen such that the four-velocity is normalized to unit Minkowski length.[4] Finally, $n_\mu(x)$ denotes the normal vector field to the hypersurface Σ, which is always timelike and future-directed [see (11.26)].

If we consider the equation in a special coordinate system in which Σ corresponds to a surface of constant time coordinate, $\Sigma = \{(s, \mathbf{x}) : s = t\}$, we have $n_\mu \equiv (1, 0, 0, 0)$ and recover the familiar $|\psi|^2$ density for the zero components of the vector field as in the non-relativistic equation (12.3). Equation (12.5) is also called the *hypersurface Bohm–Dirac equation*.

Why did we put "relativistic generalization" above in quotes? Because such a construction, with the new geometric structure of a preferred foliation, would seem to violate, if not the letter, then at least the spirit of relativity, as it basically reintroduces an absolute simultaneity through the backdoor. To appreciate this criticism, note that choosing a different foliation $\mathfrak{F}$ yields a different Bohmian theory, i.e., different trajectories! This is what people usually have in mind when they say that Bohmian mechanics cannot be made relativistic.

The latter conclusion is, however, premature. For one, such a dynamically preferred foliation of spacetime would not lead to *empirical* violations of the relativity principle. The velocity field itself cannot be measured (see Sect. 7.3), while the statistical predictions, computed from the Lorentz covariant probability measure (11.27), come out the same in every reference frame. Therefore, the hypersurfaces of "simultaneity" (if we want to call them that) cannot be empirically detected.

Secondly, the preferred foliation doesn't have to come out of nowhere. It could be determined by a Lorentz invariant law, or even generated by the wave function.[5] After all, the role of the wave function is in any case to guide the particles, i.e., determine their trajectories in spacetime. If this requires an additional structure which comes from the wave function itself, there is little to complain about. In some sense, the foliation is there anyway, within the wave function which is part of every quantum theory. The wave function determines, for instance, a Lorentz covariant energy–momentum tensor (most easily in the second-quantized formalism), which in turn distinguishes a foliation of vanishing total momentum—something like the center-of-mass frame of the universe. This may be a little (though only a little)

---

[4]This is just a convenient choice. The worldline can be parametrized arbitrarily, with the proportionality factor changing accordingly. The four-velocity only has to be parallel to the vector field on the right-hand side of the equation.

[5]For more details, see D. Dürr, S. Goldstein, T. Norsen, W. Struyve, and N. Zanghì, Can Bohmian mechanics be made relativistic? Proceedings of the Royal Society A **470**, 20130699 (2013).

reminiscent of general relativity, where the energy–momentum tensor determines the spacetime geometry via the metric; here, we get an additional geometric structure in the form of a preferred foliation.

The corresponding hypersurface Bohm–Dirac theory is Lorentz invariant. It is relativistic in the sense that massive particles move at speeds below c, and in the sense that superluminal signalling is excluded in quantum equilibrium. Finally, it is relativistic in the sense that all reference frames are equivalent for extracting empirical (statistical) predictions, so the theory can ground the measurement formalism postulated in standard relativistic quantum theories.

Remark 12.1 (Statistical Analysis of the Hypersurface Bohm–Dirac Theory)
Above, we introduced the densities $\rho_\Sigma(x_1, \dots, x_N) = n_{\mu_1}(x_1) \dots n_{\mu_N}(x_N) j_\psi^{\mu_1 \dots \mu_N}$ $(x_1, \dots, x_N)$ as the relativistic generalization of the $|\psi|^2$ density. The next step would be to give a statistical interpretation of the corresponding measure, such that

$$\mathbb{P}\big(X_i(\Sigma) \in d\sigma(x_i),\ 1 \leq i \leq N\big) = \rho_\Sigma(x_1, \dots, x_N)\, d\sigma(x_1) \dots d\sigma(x_N) \qquad (12.6)$$

yields the probability that the N particle worldlines cross the hypersurface Σ (with normal vector field n^μ) in the volume elements $d\sigma(x_1), \dots, d\sigma(x_N)$. However, we must remember the derivation of the Born rule in Bohmian mechanics: in order to interpret ρ_Σ as the equilibrium distribution of Bohmian particles, the measure must be equivariant, that is, transported with the particle dynamics.

In the relativistic formulation, this means that, if $T_\Sigma^{\Sigma'} : \Sigma^N \to \Sigma'^N$ denotes the evolution of the particle trajectories between the hypersurfaces Σ and Σ' (the relativistic flux), such that $T_\Sigma^{\Sigma'}\big(X_1(\Sigma), \dots, X_N(\Sigma)\big) = \big(X_1(\Sigma'), \dots, X_N(\Sigma')\big)$, the density would have to transform according to $\rho_{\Sigma'} = \rho_\Sigma \circ T_{\Sigma'}^{\Sigma}$. This is, in fact, equivalent to the existence of a continuity equation. But now we discover that, apart from the special case in which the wave function factorizes into a product, equivariance cannot hold in arbitrary hypersurfaces. In general, it will hold only in the preferred foliation on which the hypersurface Bohm–Dirac equation (12.5) is defined. In other words, the typical crossing probabilities for an ensemble of relativistic Bohm–Dirac particles will only correspond to ρ_Σ if the hypersurface Σ is a leaf of the preferred foliation $\mathfrak{F}$.

Why did we say then that the preferred foliation cannot be empirically detected? In other words, why does the foliation-dependence of the equivariant distribution not lead to contradictions with Lorentz invariant quantum predictions? In fact, insofar as results of measurements are concerned, the predictions of the hypersurface model are the same as those derived from the standard quantum formalism, for positions or any other quantum observables, regardless of whether or not they refer to a common hypersurface Σ belonging to $\mathfrak{F}$.[6] This is because measurement outcomes can ultimately be reduced to the orientations of instrument pointers, counter readings,

---

[6]For a detailed discussion of why that is so, we refer to K. Berndl, D. Dürr, S. Goldstein, N. Zanghì, Hypersurface Bohm–Dirac models, Phys. Rev. A **60**, 2729–2736 (1999), arXiv:quant-ph/9801070;

the ink distribution on computer printouts, and so on, i.e., records in the form of (macroscopic) particle configurations that we may as well consider in a common hypersurface Σ belonging to the preferred foliation. Since the statistics of these are correctly described by the ρ_Σ measure, and the measure transforms covariantly, they will be consistent with the predictions computed in any frame.

In all this, the non-passive character of measurements in quantum mechanics must be taken into consideration (as discussed in more detail in the second reference of footnote 6). In particular, a measurement can affect even distant particle trajectories, so that the resulting positions—and hence their subsequently measured values—are different from what they would have been if no measurement had occurred.

12.3   Nonlocality Through Retrocausality

We would like to take a step back and consider the tension between relativity and nonlocality from a more general point of view. We consider once again the scenario of the EPR experiment with correlated measurement events $\mathscr{A}$ and $\mathscr{B}$ in spacelike separated regions of spacetime (so that no signal travelling at most with the speed of light could be sent between them). In relativistic spacetime this means, in particular, that there is *no objective temporal order* between those events. In some reference frames, $\mathscr{A}$ occurs before $\mathscr{B}$, in other reference frames, $\mathscr{B}$ occurs before $\mathscr{A}$ (see Fig. 12.1). How can we nonetheless account for the fact that the two events can influence each other?

Building on a simple argument due to Gisin,[7] we can identify two assumptions that together lead to a contradiction with Bell's theorem and the empirically confirmed violations of Bell's inequality:

I. All relativistic reference frames are equivalent for the prediction of (probabilities of) measurement outcomes.
II. In every reference frame, the predictions are independent of future events.

How can we see that these assumptions lead to a violation of Bell's theorem? Suppose both I and II are satisfied. As in Chap. 10, we denote the results of spin measurements in freely chosen directions a and b by $A, B \in \{\pm 1\}$, respectively. Our measurement events are thus $\mathscr{A} = (A, a)$ and $\mathscr{B} = (B, b)$. First, we describe the experiment in a reference frame in which measurement $\mathscr{A}$ occurs before $\mathscr{B}$. Then, by assumption II, the outcome probabilities predicted for A are independent

EPR-Bell nonlocality, Lorentz invariance, and Bohmian quantum theory, Phys. Rev. A **53**, 2062–2073 (1996), arXiv:quant-ph/9510027.

[7]N. Gisin, Impossibility of covariant deterministic nonlocal hidden-variable extensions of quantum theory. Physical Review A **83**, 020102 (2011).

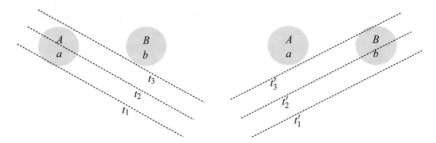

Fig. 12.1 Spacetime diagram of the EPR experiment in two different Minkowski frames. In the first frame (*left*), $\mathscr{A} = (A, a)$ occurs before $\mathscr{B} = (B, b)$, and in the second (*right*) (B, b) occurs before (A, a)

of (B, b), since the result B and the choice of b occur at later times. We denote the prediction of our candidate theory by

$$\mathbb{P}_1(A \mid B, b, a, \lambda_1) = \mathbb{P}_1(A \mid a, \lambda_1),$$

where, following the notation from Chap. 10 , λ_1 encodes all variables that could be relevant for the prediction, but are statistically independent of a and b. Notably, the argument also applies to *deterministic* theories à la Bohm, in which case the predicted outcome "probabilities" would always be 1 or 0 (assuming that λ_1 is a complete description of the initial state).

According to assumption I, however, we can equally well describe the experiment in a different reference frame in which $\mathscr{B}$ occurs prior to $\mathscr{A}$. Then, according to II, the predictions for B will be independent of (A, a), that is,

$$\mathbb{P}_2(B \mid A, a, b, \lambda_2) = \mathbb{P}_2(B \mid b, \lambda_2).$$

But put together with $\lambda = \lambda_1 \cup \lambda_2$, this defines a *local* model [see (10.12)], and we know that this cannot make the correct statistical predictions, i.e., it cannot violate the Bell/CHSH inequality, after averaging over λ.

The hypersurface Bohm–Dirac theory explicitly negates assumption I by assuming a preferred foliation of spacetime along which the guiding equation has to be evaluated. In other words, it accounts for nonlocal influences by denying the fundamental equivalence of all Lorentz frames.

We should be aware of another logical possibility, namely giving up assumption II and admitting that the theoretical predictions for certain events might indeed depend on what happens in their future. In a relativistic theory, such *retrocausal* influences could be realized by *advanced* interactions. These are interactions along past light cones, as opposed to *retarded* interactions, which "propagate", so to speak, at light speed into the future. The appeal of such retrocausal models is that, in contrast to a preferred foliation into simultaneity hypersurfaces, the light cone structure is a genuine part of relativistic spacetime geometry. In fact, if it is assumed that the fundamental laws are time-symmetric, making no a priori distinction between past and future, it is only natural to expect advanced and retarded effects in relativistic theories. Just think of classical electrodynamics, where Maxwell's equations have both advanced and retarded solutions. There, the advanced solutions are usually dismissed out of hand as "unphysical", but it is actually a very important and difficult problem to explain why we don't observe advanced radiation. This issue is intimately linked with the emergence of irreversible behavior and the second law of thermodynamics, which we mentioned very briefly in Sect. 3.5.3.

In any case, one possible idea is that retrocausal effects come into play in relativistic quantum mechanics to account for nonlocal correlations. (Of course, we would then have to explain why we don't experience retrocausality on macroscopic scales.) There are some speculative models working on this ansatz, including one by Goldstein and Tumulka[8] that evaluates the Bohmian velocity field along future light cones. Other models postulate two types of wave function: retarded waves propagating from past to future, and advanced waves propagating from future to past.[9] The biggest problem with retrocausal theories is, in general, their statistical analysis. It is not clear how to find an equivariant measure for an advanced guiding law, or how to formulate a statistical hypothesis about wave functions "from the future". For this reason, there is no fully-developed retrocausal theory that could make serious claims to grounding a relativistic quantum formalism. But as we said, it is one possible avenue to pursue.

We should note that denying assumption II does not necessarily imply explicit retrocausation in the sense of advanced interactions. Strictly speaking, our argument referred only to "future events" relative to a given time coordinate, not the causal (light cone) structure of relativistic spacetime. Moreover, if we consider fundamentally stochastic theories, the status of assumption II is already less clear. Should we be asking whether conditioning on future events affects the probability of earlier ones? Or is it more like saying something like: "The fact that the street is wet tomorrow morning increases the probability that it will rain tonight"?

In any case, we will soon discuss the relativistic generalization of the stochastic GRW theory, and it is interesting to reflect on the causal structure of this theory—or rather, the lack thereof. Technically, the theory violates assumption II by allowing the probabilities of the measurement outcomes A and B to depend nonlocally on distant collapse events (flashes). Hence, relative to one frame, the wave function may have already collapsed, while relative to another, it has not yet collapsed, but nonetheless, no serious worries arise about the future influencing the past. In fact, it doesn't make sense to ask whether A has caused the spacelike separated event B or

---

[8]S. Goldstein and R. Tumulka, Opposite arrows of time can reconcile relativity and nonlocality. Classical and Quantum Gravity **20** (3), 557–564 (2003).

[9]See, e.g., R.I. Sutherland, Causally symmetric Bohm model, Studies in History and Philosophy of Modern Physics **39** (4), 782–805 (2008). See also B. Reznik and A. Aharonov, On a time symmetric formulation of quantum mechanics, Physical Review A **52**, 2538–2550 (1995).

vice versa; all that the theory describes is a consistent Lorentz invariant probability law for the appearance of flashes.

12.4   Bohmian "Big Bang" Model

As something of a side note, we will now present a Bohmian toy model which is sometimes referred to as the Bohmian "big bang" model. We call it a toy model because it is not a serious candidate for a fundamental theory of Nature. The model is nonetheless interesting and instructive. For one, it shows that there is indeed no fundamental contradiction between relativity and nonlocality—not even in a deterministic theory of particles. On the other hand, it is a good preparation for our examination of the relativistic GRW theory, which picks up a crucial "trick" from this model.

To define it, we assume a distinguished point $\mathscr{O}$ in Minkowski spacetime from which all particle trajectories originate. This means, in particular, that all particle trajectories lie in the forward light cone of $\mathscr{O}$, that is, in the region

$$\mathscr{M}_0 = \left\{ x^\nu \mid x^\mu x_\mu > 0, \ x^0 > 0 \right\},$$

if we choose $\mathscr{O}$ as the origin of our coordinate system. We can thus think of $\mathscr{M}_0$ as all there is, i.e., the entire spacetime manifold (a geometry also know as the *Milne model*). For obvious reasons, the event $\mathscr{O}$ is referred to as the "big bang", but our toy model should not be confused with the actual Big Bang cosmology since it is empirically inadequate in many ways. In particular, our actual universe is not confined to a single light cone or *event horizon*.

In any case, the point of this model is that our truncated spacetime $\mathscr{M}_0$ allows for a natural Lorentz invariant foliation defined solely in terms of the spacetime metric, viz.,

$$\mathscr{M}_0 = (\Sigma_s)_{s>0}, \quad \Sigma_s := \left\{ x \in \mathbb{R}^4 : |x| = \sqrt{x^\mu x_\mu} = s, \ x^0 > 0 \right\}. \tag{12.7}$$

The leaves Σ_s of the foliation are thus hyperboloids with constant Minkowski distance s from the origin, as shown in Fig. 12.2.

More technically speaking, these hyperboloids are equipotential surfaces (surfaces of constant value) for the function $\varphi(x) = \sqrt{x^\mu x_\mu}$. The corresponding normal vector field is parallel to the gradient of $\varphi(x)$, that is,

$$n^\mu(x) = \partial^\mu \varphi(x) = x^\mu / |x|.$$

It is easy to check that this normal vector field is timelike and future-directed, which means that the hypersurfaces Σ_s are indeed spacelike. To make things simple, we can use the same parameter $s \in \mathbb{R}^+$ to parametrize both the leaves of the foliation and the worldlines of the particles, writing $X_k^\mu(s) := X_k^\mu(\Sigma_s)$.

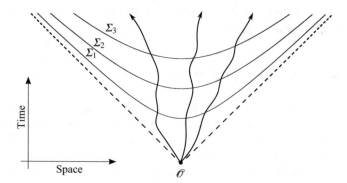

Fig. 12.2 Bohmian "big bang" model. *Dotted lines* indicate the future light cone of the "big bang" $\mathcal{O}$. The guiding equation is defined along the Lorentz invariant hyperboloids Σ_t

Analogously to (12.5), we can now define the guiding equation:

$$\frac{d}{ds} X_k^{\mu_k}(s) \propto j_\psi^{\mu_1 \cdots \mu_k \cdots \mu_N}(x_1, \ldots, x_N) \prod_{j \neq k} \frac{x_{j,\mu_j}}{|x_j|}\Bigg|_{x_j = X_j(s)}. \tag{12.8}$$

This equation is fully relativistic in that it uses only the spacetime metric to define the leaves of the foliation on which the right-hand side is evaluated. Note that, not unlike the real Big Bang, we have a singularity at $\mathcal{O}$.

12.5   Relativistic GRW Theory

Finally, we will briefly introduce a relativistic generalization of the GRW theory that goes back to the work of Roderich Tumulka.[10] We cannot go into all the technical details, but just highlight the main ideas. In the non-relativistic GRW theory, the wave function collapses instantaneously whenever a flash $(\mathbf{X}, T)$ occurs, that is (in a spatio-temporal sense) along the equal-time hyperplane $t = T$. More concretely, the wave function is multiplied at time T by a Gaussian form factor centered around $\mathbf{X}$ (this is the "spontaneous localization"), while evolving according to the linear Schrödinger equation in-between two collapse events.

In the generalization to relativistic spacetime, the wave function is a multi-time wave function which evolves freely (the model presented by Tumulka is without interactions) outside flash events. The equal-time hyperplanes are replaced by Lorentz invariant hyperboloids as just discussed in the Bohmian big bang model. In that model, however, we had to introduce in an *ad hoc* manner a distinguished event in spacetime that would specify the locus of a universal light cone in which

---

[10]R. Tumulka, A relativistic version of the Ghirardi–Rimini–Weber model, *Journal of Statistical Physics* **125** (4), 821–840 (2006).

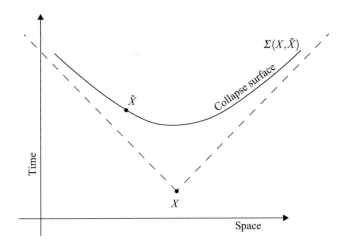

Fig. 12.3 Collapse in rGRWf. The wave function collapses along the Lorentz invariant hyperboloid of constant Minkowski distance between successive collapse events (flashes)

"time unfolds" along the hyperboloids Σ_t. In the GRW theory, we already have the flashes, i.e., the collapse events, as distinguished events in spacetime! That is, each flash defines the locus of a light cone whose interior can be foliated by Lorentz invariant hyperboloids; and of these hyperboloids, one will be chosen at random (essentially by an exponential waiting time) for the occurrence of the next flash, and as the hyperplane along which the wave function will be localized by multiplication with a Gaussian form factor.

Omitting further technical details, the relativistic GRW theory with flash ontology (rGRWf) can thus be described as follows (see Fig. 12.3):

1. Start with the wave function on an arbitrary spacelike hypersurface Σ_0, and a set of initial "seed flashes" $X_1, X_2, \ldots, X_N \in \mathcal{M}$, with one flash for each particle degree of freedom.
2. Between two flashes, the (multi-time) wave function evolves according to the (free) unitary time evolution. The wave function also defines a probability distribution on spacetime for the occurrence of subsequent flashes. For each "particle", a new flash X is always timelike separated from the preceding one $\tilde{X}$.
3. If such a flash event occurs, let's say for the i th particle at spacetime point $\tilde{X}_i$, the wave function collapses along the Lorentz invariant hyperboloid

$$\Sigma(X_i, \tilde{X}_i) = \left\{ y \in \mathcal{M} : |y - X_i| = |\tilde{X}_i - X_i|, \ (y - X_i)^0 > 0 \right\}.$$

4. We thus obtain a new generation of flashes $\tilde{X}_1, \tilde{X}_2, \ldots, \tilde{X}_N \in \mathcal{M}$, and a new collapsed wave function. With these, we can repeat the procedure to obtain the next generation of flashes, and so on.

By construction, this iterated process yields a Lorentz invariant probability distribution for the flashes (which can also be shown to be independent of the separation of flashes into "generations"). We must therefore take the rGRWf theory seriously as a theory *about the flashes*—conceived, ontologically, as discrete matter points— in which the role of the wave function is to define the stochastic law for their appearance.

Epilogue

13

> Thus shall thou learn all things: the unshaken heart of the well-rounded truth, as well as the apparent truth conceived by mortals which is without faith and without truth. But yet thou shalt learn also how the apparent truth should receive validity and penetrate the world view completely.
>
> Parmenides, On Nature[1]

In this book, we have discussed three quantum theories that solve the measurement problem and ground the statistical predictions of quantum mechanics in an objective and coherent description of Nature. Now it's only natural to ask which of these theories is actually true. Why did our quest for an understanding of quantum mechanics lead us to three competing formulations rather than just one "final" version?

We chose the measurement problem as the starting point of our investigation, and this led to three possible solutions that are realized in Bohmian mechanics, GRW, and the Many Worlds theory, respectively. Of course, this does not necessarily mean that all three solutions are equally compelling. We don't want to hide the fact (and careful readers have probably already noticed) that we have a preference for Bohmian mechanics. For us, quantum mechanics *is* Bohmian mechanics. The goal of this book, however, was not to convince readers of this view but to provide the necessary foundation that allows them to make their own judgement.

In science, we are always in the situation that observable phenomena can merely constrain but not conclusively determine the "correct" theory. When it comes to quantum physics, this is aggravated by the fact that our epistemic access to the microscopic state of affairs is limited, in principle, and we have striking illustrations of the fact that different theories can provide radically different descriptions of the world even when they are empirically equivalent. This is why we have to resort to criteria such as beauty, simplicity, and explanatory power, criteria that

---

[1] Authors' translation.

© Springer Nature Switzerland AG 2020
D. Dürr, D. Lazarovici, *Understanding Quantum Mechanics*,
https://doi.org/10.1007/978-3-030-40068-2_13

have a subjective quality and need not lead to a consensus. Occam's razor—the philosophical principle that "entities should not be multiplied without necessity" or, more generally, that one should always prefer the most parsimonious of all possible explanations—can be turned just as well against the Bohmian trajectories as against the additional parameters of GRW, or the Many Worlds of the Everett interpretation. Thus, in the end, the razor remains rather blunt.

In a spirit of reconciliation, it should be noted, however, that Bohm, GRW, and Everett do, in fact, agree on many important points. And appreciating their common message goes a long way towards a coherent understanding of quantum mechanics.

All three theories are "quantum theories without the observer", which is to say that they develop an objective description of Nature in which "observers" or "measurements" have no a priori distinguished status (Chap. 2).

All three theories involve some form of nonlocality, which we know they have to—as we learned from Bell's theorem—in order to account for observed phenomena (Chap. 10).

All three theories agree on the fact that observable operators are not fundamental, but rather convenient mathematical tools for statistical bookkeeping (Chap. 7). They also explain why "observable values" do not generally reflect pre-existing properties of the measured system, but are rather "produced" as a result of the measurement process (Chap. 9).

All three theories compel us to take the wave function on configuration space seriously (Chap. 1). This means, on the one hand, that the "position representation" of the wave function is distinguished, which should not be surprising given that, at the end of the day, we always have to relate it in one way or another to physical objects located in three-dimensional space. And it means, in particular, that the wave function is not just "information" or statistical bookkeeping, but real physical degrees of freedom that feature in the objective state description of physical systems. The wave function is affected by measurement or "preparation" processes. The wave function establishes a real physical connection between entangled systems. And the wave function produces the measurement outcomes, so to speak, rather than just describing their statistics or our incomplete knowledge of pre-existing values. There is a big philosophical debate about the exact metaphysical status of the wave function,[2] but it is absolutely impossible to understand quantum mechanics without acknowledging its active physical role.[3]

A notable point of disagreement between the three quantum theories concerns the status of randomness and probabilities (Chap. 3). It is particularly significant because of the common wisdom that regards randomness or indeterminism as *the* key innovation of quantum mechanics. In fact, both Bohmian mechanics and the

---

[2]See, e.g., A. Ney and D.Z. Albert (eds.), *The Wave Function: Essays on the Metaphysics of Quantum Mechanics*. Oxford University Press, 2013.

[3]A much discussed attempt to substantiate the "reality" of the wave function with a rigorous mathematical proof is the PBR theorem: M.F. Pusey, J. Barrett, and T. Rudolph, On the reality of the quantum state. Nature Physics **8**, 475–478 (2012).

Many Worlds theory are deterministic. Bohmian mechanics involves "randomness" in the sense of in-principle *unpredictability*, but not in the sense of fundamental *indeterminism*. Collapse theories such as GRW are indeed indeterministic, i.e., involve real irreducible randomness. Somewhat ironically, this is also the class of theories whose predictions deviate from those of standard quantum mechanics. Hence, one could say that, if quantum mechanics is really indeterministic, it is not exact. And if quantum mechanics is exact, it is not indeterministic.

In a nutshell, the current state of quantum foundations is the following. After clearing away the rubble of the past century, one is basically left with three serious candidates for a precise formulation of non-relativistic quantum mechanics. Experiments distinguishing between theories with and without spontaneous collapse are possible in principle, and it seems to be only a matter of time until they will be carried out in practice. If we found empirical evidence for spontaneous collapse, then GRW, or other continuous collapse theories into which the GRW ansatz has been developed, would win the day. At the same time, it would certainly increase interest in the question as to whether the stochastic collapse mechanics is indeed a fundamental law of Nature or rather an effective description of a more fundamental theory yet to be discovered.

If the superposition principle for the wave function is universally valid, Bohmian mechanics and Many Worlds are the best available alternatives. Bohmian mechanics can be understood to the point that no serious foundational issues remain open in the context of non-relativistic quantum mechanics. It may be possible to arrive at an equally deep understanding of the Many Worlds theory, but the authors have not yet been able to do so. The discussion provided in this book reflects our best efforts and this may help some readers to progress further.

To end with, we would advise physics students against spending the next hundred years on foundational debates about non-relativistic quantum mechanics. Today, the most important open problems concern relativistic quantum physics. And one of the great achievements of all the precise non-relativistic theories is to point us to the right questions and possible paths for future progress.

Index

© Springer Nature Switzerland AG 2020
D. Dürr, D. Lazarovici, *Understanding Quantum Mechanics*,
https://doi.org/10.1007/978-3-030-40068-2

Printed in the United States
by Baker & Taylor Publisher Services